Computational Algorithms for Closed Queueing Networks

THE COMPUTER SCIENCE LIBRARY

Operating and Programming Systems Series
PETER J. DENNING, *Editor*

1	Halstead	A Laboratory Manual for Compiler and Operating System Implementation
2	Spirn	Program Behavior: Models and Measurements
3	Halstead	Elements of Software Science
4	Franta	The Process View of Simulation
5	Organick and Hinds	Interpreting Machines
6	Wilkes and Needham	The Cambridge CAP Computer and Its Operating System
7	Bruell and Balbo	Computational Algorithms for Closed Queueing Networks

OPERATING AND PROGRAMMING SYSTEMS SERIES

7

Computational Algorithms for Closed Queueing Networks

Steven C. Bruell
University of Minnesota

Gianfranco Balbo
University of Torino

NORTH HOLLAND
NEW YORK • OXFORD

Elsevier North Holland, Inc.
52 Vanderbilt Avenue, New York, New York 10017

Distributors outside the United States and Canada:

Thomond Books
(A Division of Elsevier/North-Holland Scientific Publishers, Ltd.)
P.O. Box 85
Limerick, Ireland

Library of Congress Cataloging in Publication Data

Bruell, Steven C
Computational algorithms for closed queueing networks.
(Operating and programming systems series; v. 7)
(the Computer science library)

Bibliography: p.
1. Queueing theory. 2. Algorithms. I. Balbo, Gianfranco, joint author. II. Title.
T57.9.B78 519.8′2 80-15754
ISBN 0-444-00421-1

Manufactured in the United States of America

to Sandy and Laura

TABLE OF CONTENTS

PREFACE

This book deals with the analysis of single and multiple job class queueing network models. Starting from the steady state distribution of jobs in such networks, a number of computationally efficient algorithms are developed to numerically evaluate performance measures of interest. Two methods are discussed in detail: the normalization constant approach and the mean value analysis approach. Several practical problems in the implementation of the computational algorithms under each scheme are discussed.

The book is a condensation of the first authors Ph.D. thesis at Purdue University. The algorithms described in the text have all been implemented (and tested) in a program we call PNET -- the Purdue queueing network evaluator.

We intend this book to be useful to professionals and/or researchers in the performance evaluation field as well as serious students who wish to gain a knowledge of queueing networks without having to plow through the literature.

We especially thank Professors Peter J. Denning and Herbert D. Schwetman both of Purdue University for their constant support and guidance during the course of writing this book. The content of this book was developed while the authors were in the Computer Sciences Department of Purdue University.

We also thank Professor Saul Rosen, director of the Purdue University Computing Center, and the staff of the computing center for providing a computing facility that made the development and preparation of this book possible.

S. C. Bruell, University of Minnesota
G. Balbo, University of Torino

LIST OF TABLES

LIST OF FIGURES

CHAPTER 1

INTRODUCTION

1.1 MOTIVATION

The number of large scale computer systems has increased rapidly with recent technological advances. More powerful processing elements and less expensive fast-access memory have led to the development of complex computer systems with such features as multiprogramming, multiprocessing, virtual memory, and interactive processing. As the complexity of computer systems increases, a thorough understanding of the performance of these systems exceeds the limits of intuition.

There have been many cases in which sophisticated systems have been designed without a complete understanding of the characteristics of the future workload, and of the interactions among their component modules. The system's performance fell short of the expectations, providing computing capabilities far below the apparent power of the equipment. Bottleneck analysis has sometimes pointed out, in retrospect, the existence of design problems that are not easy to fix.

These sorts of failures constitute strong motivations for studying modeling and performance evaluation of computer systems. Complex systems cannot be understood without suitable tools. The high cost of computer systems and the irrevocable nature of certain design decisions require an analyst to determine, with reasonable accuracy, the performance of new systems even before building the first prototype.

Performance evaluation is not just determining whether or not a system meets certain objectives; it is also understanding if and how system performance can be improved. Models of computer systems are used as an aid in understanding system performance. It is useful to formulate a model which both describes the system and predicts its behavior as its components and parameters are changed. Baskett points out that models of computer systems are useful not only for predicting performance but also as a framework for making and interpreting measurements [BASK73].

Models of computer systems can thus be considered as the basic tools by which the computer analyst understands the behavior of a given system. They are also used for calculation, consistency checking and prediction of performance figures [DENN78b].

Models must be able to accurately represent the behavior of computing systems. To assess a model's accuracy, we must compare its calculated performance figures with those measured on the real system, over a complete set of independent experiments. This is called validation of a model.

A validated model can then be used to predict the performance of hypothetical systems which are modifications of the original one.

Figure 1-1 repesents a schema for validation. A workload is run on a real system and, over a given observation period, performance figures and model parameters are measured [DENN78b]. The input parameters of the model are used to calculate performance quantities which are then compared with the ones actually measured. If good agreement is observed over many different observation periods we come to believe that the model is capable of reproducing the behavior of the system.

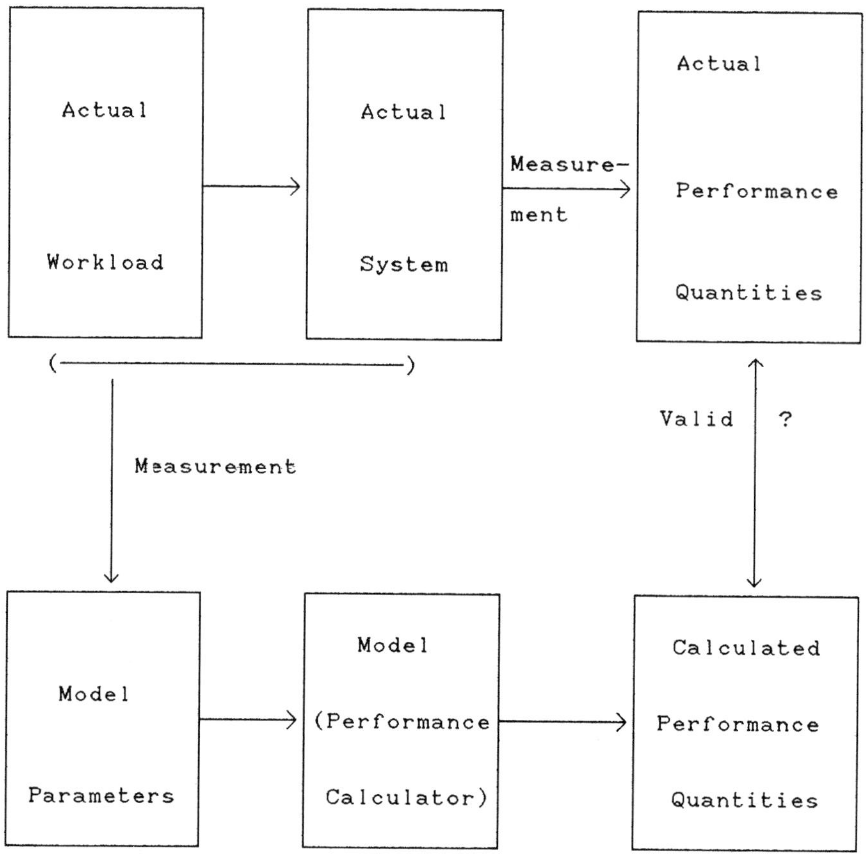

Figure 1-1: Typical validation schema

Figure 1-2 represents the schema usually followed during a prediction study. An analysis of the projected modification of the workload, or of the projected modification of the system, or of both, can be used to derive a model of the projected configuration. Its evaluation produces a set of projected (forecasted) performance figures. If and when such modifications are implemented, a validation study of the model can be conducted again.

Validating a model is actually more complex than this general discussion might suggest. Statistical techniques should be used to test the significance of results. The models should be analyzed for sensitivity to input parameters in order to establish whether they are stable with respect to the parameters whose values must be forecast, and to allow the construction of confidence intervals for the computed performance figures.

As stated previously, computer performance analysts typically use models for three purposes: calculation, prediction of performance measures, and consistency checking [DENN78b]. These models fall into two major categories: simulation and analytic.

The advantage of using simulation models is that they can achieve any degree of accuracy desired in reflecting the

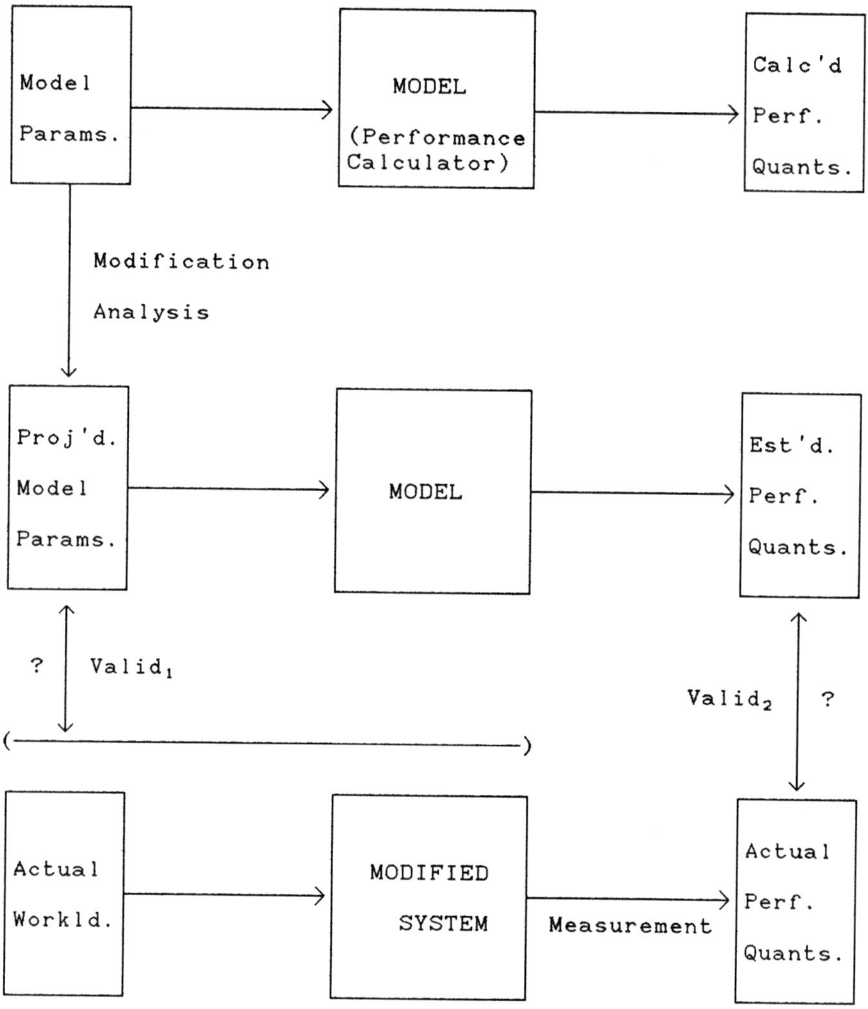

Figure 1-2: Prediction schema

detailed structure of a system. This advantage over analytic models explains why simulation is the more popular of the two modeling approaches. But this advantage can also be deceptive, because the level of detail in a simulation model can be considerable. As more detail is included in the model, simulation becomes more costly. This can severely limit the utility of simulation models.

Some of these problems do not arise with analytic models. Many analytic models admit closed form solutions or efficient computational procedures which provide insight into the mechanisms underlying a system.

Queueing network models, in particular, are well suited as computer system models because they are able to capture the interaction between a system's workload and a system's resources. The model treats a computer system as a network of processors (CPU's, I/O devices, terminals, etc.) and a collection of customers (jobs, tasks). Queueing delays may occur for any system resource which can be used by at most a few jobs at a time. These models can be used to locate performance problems in a system (e.g. bottlenecks). They have also proved to be accurate estimators of some system performance measures such as resource utilizations.

This result is somewhat surprising since many of assumptions underlying these queueing network models are violated by real systems. In a recent series of publications, Buzen and Denning, using an operational approach, have explained this phenomenon [BUZE76a,b,c,d, DENN77, DENN78b]. In section 1.3 we discuss these results further.

1.2 PLAN OF BOOK

The primary objective of this book is to describe efficient computational algorithms for evaluating the performance measures of closed queueing network models. In Chapter 2 we extend the computational algorithms developed by Buzen [BUZE71b,BUZE73] to queueing networks with multiple job classes. Our methods can handle networks in which jobs may switch from one job class to another while traversing the queueing network.

Chapter 3 deals with some practical considerations arising from the solution of the single and multiple job class queueing network models developed in Chapter 2. First, we discuss a method for restructuring some queueing networks to reduce the computational effort in evaluating the original network's performance measures. Second, we derive a simple means of controlling arithmetic overflow and/or underflow in the computational algorithms. Finally, we consider the implementation of these algorithms.

In Chapter 4 we study the same class of queueing network models, but from a slightly different point of view. We show that this approach, called mean value analysis, is quite simple and leads to simpler algorithms for evaluating the performance measures of single and multiple job class queueing network models [REIS78].

Chapter 5 summarizes the contents of the book. We also make recommendations when to use the algorithms described in Chapter 2 and when it is better to use the algorithms of Chapter 4 for computing the performance measures of queueing network models.

The remainder of this chapter provides important background information on queueing network models of computer systems.

1.3 QUEUEING NETWORK MODELS OF COMPUTER SYSTEMS

The behavior of a computer system depends on the interaction of the system workload and the system resources. The workload is characterized by the service requirement of jobs for the system resources. Queueing networks capture this interaction in a conceptually clear model. The models are used for calculating and predicting performance statistics such as resource utilizations, throughputs, and average queue lengths. They are also useful in determining where jobs are likely to encounter bottlenecks and how the characteristics of one resource can affect other parts of the system. This section reviews the basic terminology and the main results for these models.

A queueing network is a collection of stations (or service centers) arranged in such a way that customers proceed from one to another in order to fulfill their service requirements. In applying queueing networks to computer systems, the stations represent the various system resources (e.g. CPU's, channels, disks, and drums) while the customers represent jobs in the system. In the following discussion we will consistently employ this computer related terminology.

Each station has an associated queue in which jobs may wait prior to receiving service. The stations are characterized by a service rate and a queueing discipline. The service rate is expressed as the number of jobs departing the station per unit busy time of the station. This service rate may be load dependent, i.e., the rate at which the station processes work may depend on the length of the associated queue. The queueing discipline determines the order in which arriving jobs receive service. The simplest discipline processes jobs in order of arrival, i.e., first come first served (FCFS).

Early queueing analysis of computing systems used classical queueing theory, which treated the system as a single queue serving an infinite population of jobs (see Figure 1-3). This model does not allow the analyst to represent the internal structure of the computing system: it gives no clues to the relation between external behavior and internal configuration. Moreover, this model cannot represent the common situation of a "finite population" of users; typically, the arrival rate of new jobs decreases when the total number of jobs in the system increases.

To overcome the "infinite population" limitation, the model of Figure 1-4 was chosen to describe interactive systems. This model represents a computer system with N terminals, one for each user logged on. The number of jobs circulating throughout the system is N and each job can either be in one of three states: (1) "thinking" at a terminal; (2) waiting in the queue to enter the central system; or (3) being served by the central system.

This model is known in classical queueing theory as the "machine repairman" model. Scherr applied it to the Compatible Time Sharing System (CTSS) at MIT; he was able to estimate the mean response time of the system with surprising accuracy [SCHE67].

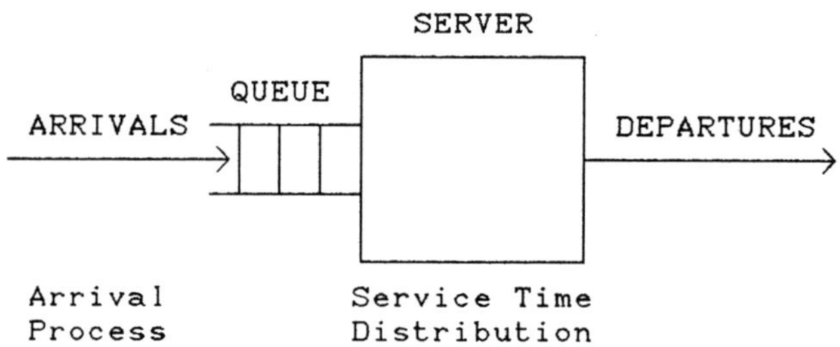

Figure 1-3: A single resource queue

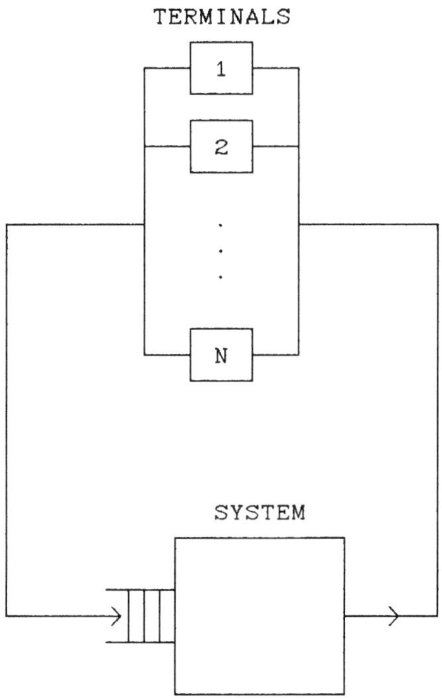

Figure 1-4: A finite population model

The structure of the model of Figure 1-4, however, fails to represent a contemporary computer system. CTSS was a swapping system -- only one job at a time ran -- and so the single-resource model of the central subsystem was adequate. More complex models are needed to represent multiprogrammed and multiprocessing computer systems where several jobs can execute simultaneously.

A significant breakthrough occurred when Jackson developed a solution technique for a special class of queueing networks [JACK57]. Jackson considered the analysis of open networks of queues with a single class of jobs. A network is considered open if jobs are permitted to enter or leave, and closed if the same jobs remain in the network at all times. Operationally, this implies that we use one mean and one set of routing probabilities to describe all jobs. No attempt is made to specify separate behaviors for different groups of jobs.

Each station in Jackson's model contains one or more parallel servers. Service time distributions at the various stations are assumed to be exponential. A state in this network model represents an apportionment of jobs among stations. For an M station network, a state is a vector $\underline{n}=(n_1,n_2,...,n_M)$ in which $n_i \geqq 0$ is the number of jobs at station i.

In analyzing a queueing network model, we seek a closed form expression for the equilibrium probability of finding the network in any given state $\underline{n}$; this probability is denoted by $P(\underline{n})=P(n_1,n_2,\ldots,n_M)$. Station utilizations, average queue lengths, and throughputs can be derived from these equilibrium state probabilities. Let $p_i(n)$ denote the marginal probability of finding n jobs at the ith station. Then

$$p_i(n) = \sum_{\substack{\text{all feasible states} \\ \underline{n} \text{ such that } n_i=n}} P(n_1,n_2,\ldots,n_i,\ldots,n_M).$$

The utilization, which measures the probability that a station is busy, is computed as

$$U_i = 1 - p_i(0).$$

Jackson proved that the equilibrium probability of state $\underline{n}$ in an open network factors into the product of each of the marginal distributions, i.e.,

$$P(n_1,n_2,\ldots,n_M) = p_1(n_1) \cdot p_2(n_2) \cdots p_M(n_M).$$

This is termed a "product form" since it separates into a product of factors, one factor for each station in the network.

Gordon and Newell obtained the equilibrium state probabilities for closed queueing networks with each station having an exponential service time distribution [GORD67a]. They studied a closed network with a finite number of jobs, N, requesting and receiving service from a finite number of stations, M. Each station, i, may have a_i parallel identical servers with exponentially distributed service times. A job completing service at station i will proceed directly to station j with a transition probability p_{ij} which is independent of the state of the network. The state of the network is again described by a vector $\underline{n}=(n_1,n_2,...,n_M)$ but now one additional condition is imposed, i.e., the sum of all the n_i must be N (since the network is closed).

Gordon and Newell proved that the equilibrium probability of state $\underline{n}$ also satisfies a product form:

$$P(\underline{n}) = P(n_1,n_2,...,n_M) = \frac{1}{G(N)} \prod_{i=1}^{M} f_i(n_i)$$

where G(N) is a normalization constant chosen to make all the feasible state probabilities sum to one. That is,

$$G(N) = \sum_{\substack{\text{all feasible} \\ \text{states } \underline{n}}} \prod_{i=1}^{M} f_i(n_i).$$

The function $f_i(n_i)$ depends on the characteristics of the ith station. We defer a precise description of the f_i until Chapter 2.

It is interesting to note that Gordon and Newell were unaware of Jackson's earlier work on queueing networks. They acknowledged later that their formulae for the equilibrium state probabilities could be obtained by specializing the parameters of Jackson's more general model [GORD67b]. This notwithstanding, most researchers involved in the early development of queueing network models made use of the simplified notation and results of Gordon and Newell's paper.

Though these product form solutions are easily expressed, direct evaluation of the state probabilities can be computationally expensive due to the large number of operations required to evaluate the normalization constant G(N). Even when such results were calculated, the large number of finite precision arithmetic operations involved casts doubt upon the accuracy of the results.

In the early 1970's Buzen developed an efficient computational algorithm for evaluating the normalization constant [BUZE71b,BUZE73]. Buzen also studied methods for obtaining the performance measures of a queueing network as simple functions of the model parameters and the normalization constant(s). Many real systems, thought to be too complex for efficient equilibrium analysis, could now be effectively studied using Buzen's algorithms.

After deriving these algorithms, Buzen considered methods of constructing queueing network models of computer systems [BUZE71b]. Such a model must accurately represent the interaction and movement of jobs through a computer system.

With this in mind, Buzen designed the central server model of a multiprogramming system. Figure 1-5 depicts this model. The rectangles in the diagram represent processors in the real system. Jobs (or programs) move through this network in the direction of the arrows, visiting one station at a time. Thus, in a central server model a job is represented as an alternating sequence of CPU and I/O processing intervals. If a job arrives to a station which is busy processing some other request, a queue of delayed requests forms. The queues are represented by the open ended ladders attached to each rectangle. Buzen applied these models to determine system bottlenecks [BUZE71a].

The queueing networks studied by Jackson and Gordon and Newell have two principal limitations: all jobs belong to one class, and all service time distributions are exponential. But, in a real system, each job has its own service parameters for each station, and real service time distributions are seldom exponential.

Several authors have extended queueing models to more accurately represent real computer systems [FERD71, POSN68, BASK72]. These separate models are special cases of the model developed by Baskett, Chandy, Muntz, and Palacios [BASK75].

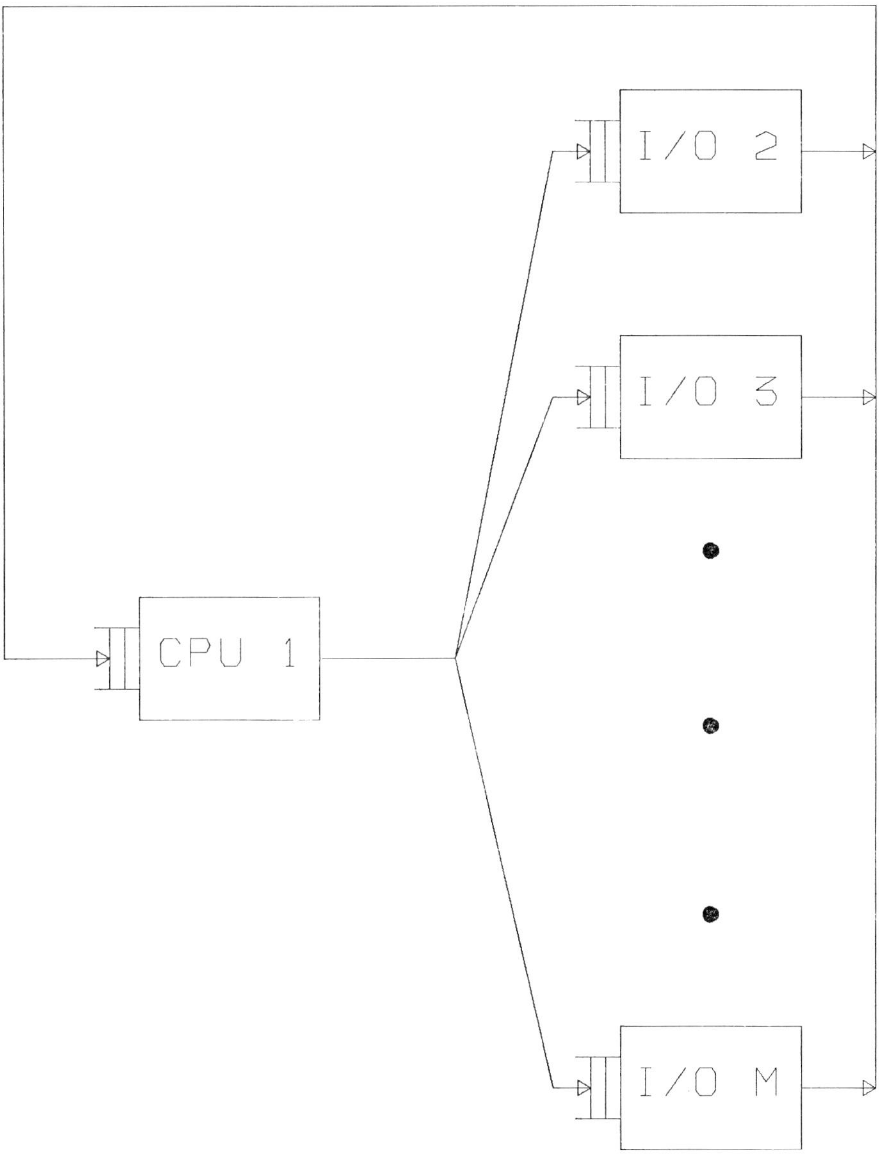

Figure 1-5:
Central Server Model

The general model described in [BASK75] relaxes the two principal limitations of the previous models. The model divides jobs into R distinct classes. Each class may have its own set of transition frequencies. Jobs can also change class according to these transition frequencies. A typical transition frequency has the form $p_{(i,r)(j,s)}$, describing the fraction of departures from station i in class r that go next to station j in class s. A state now represents a distribution of jobs over classes and stations. For an M station network, a state is denoted by a vector $\underline{\underline{n}}=(\underline{n}_1,\underline{n}_2,\ldots,\underline{n}_M)$ where each component of $\underline{\underline{n}}$ is also a vector. These vectors $\underline{n}_i=(n_{i1},n_{i2},\ldots,n_{iR})$ denote the number of class r jobs at station i (n_{ir}). We will henceforth refer to these queueing networks as BCMP networks (or models).

The BCMP model describes either an open, closed, or mixed queueing network. In a closed system a constant number of jobs, N, circulate in the network at all times. In an open system two general types of state dependent Poisson arrival processes are allowed. A mixed system is closed with respect to some classes of jobs and open with respect to other classes.

Earlier queueing network models assumed a first come first served queueing discipline at each station. In a BCMP model each service center can have any one of four queueing disciplines. For a station with a first come first served (FCFS) queueing discipline, each job has the same exponential service time distribution. The station's service rate can be load dependent, the rate only being a function of the total number of jobs at the station and not the number of jobs of any class. A station may have a single server with a processor sharing (PS) queueing discipline. The server is shared among all jobs infinitely fast in a round robin fashion. Each class of job may have a distinct service time distribution. A station where the number of servers is always at least the maximum number of jobs at the station is said to have an infinite server (IS) queueing discipline. Thus, no job will ever experience a queueing delay waiting for a server. This type of queueing discipline is typically used to represent user terminals in a time-sharing model. Each class of job may have a distinct service time distribution. Finally, a station may have a single server with a last come first served preemptive resume (LCFS) queueing discipline. That is, the most recent job to arrive is served first, and the job currently receiving service (if any) is preempted. Again, each class of job may have a distinct service time distribution.

The BCMP model is capable of accurately representing a wider range of computer systems than can the simpler, more restrictive models of Jackson and Gordon and Newell.

However, the computational complexity of obtaining performance measures in the general model may be prohibitive. In Chapters 2 through 4 we present a thorough analysis of this problem.

In a recent set of publications, Buzen and Denning have studied the assumptions necessary to solve queueing network models [BUZE76a,b,c,d, DENN77, DENN78b]. The solution of a queueing network requires that the flow into and out of system states be balanced. The global balance equations equate the total flow into and the total flow out of each system state. The large number of system states preclude a direct solution of these equations. Two assumptions are made to simplify them. The first is "one-step transitions" which asserts that multiple job arrivals and departures will not be observed. This assumption implies that the rate at which the system enters or leaves a state depends only on the rate of job flow between stations. The second assumption is "homogeneity", which asserts that flows out of a station may depend only on the station's queue length and not on how jobs are distributed elsewhere in the network. Under these assumptions, the global balance equations decompose into a large number of simpler local balance equations [CHAN72, KLEI75, SIMO79]. Any solution of the local balance equations will also be a solution of the global balance equations (but not necessarily conversely).

Locally balanced networks give rise to state probabilities which have a product form.

There are systems which do not obey local balance relations. These systems may be shown to involve some form of blocking -- i.e., one station can cause another to stop processing. This can occur if there are resource conflicts, for example, if a processor is stopped to await the release of buffer space controlled by another processor.

Of more practical importance the work of Denning and Buzen takes a positive step in the direction of an operational theory of system performance [BUZE76a,b,c,d, DENN77, DENN78b]. In their work, Denning and Buzen abstract a set of relationships between quantities which are independent of underlying model or device structures. Operational "laws" establish relationships between throughput, response time, utilization, and various other factors related to system performance. "Except for measurement errors, the operational laws apply with complete precision to all collections of observational data" [BUZE76c].

1.4 NON-PRODUCT FORM QUEUEING NETWORKS

Muntz and others believe that the methods for developing product form solutions of queueing network models have reached their limits [MUNT75, MUNT78]. There are two principal techniques for dealing with queueing networks that have no product form solution: decomposition and diffusion.

Traditionally, queueing problems are treated as continuous-time, discrete-state Markov processes. This view leads to a large system of linear equations equating the flows into and out of system states. These equations can, in principle, be solved numerically for the proportion of time each system state is occupied. However it is impractical to program the solution of these equations, for most real systems have enormous state spaces. Even for small systems the usual numerical technique, Gauss-Seidel elimination, requires considerable computation before converging to a solution.

The diffusion approximation attempts to reduce the complexity of the balance equations by regarding the individual queue sizes $n_i(t)$ as continuous random variables and transforming the balance equations to a second-order differential equation (the "diffusion equation") [KLEI75, MUNT75, COX65]. This equation can, in principle, be solved numerically - again a time-consuming procedure. In some cases a closed form solution can be found.

The interest in the diffusion technique arises from its ability to handle general service time distributions at stations with FCFS queueing disciplines. Despite its additional modeling "power", the technique has proved tedious to use in specific cases. It has been applied to study delays at terminals [GAVE71] and CPU utilization in multiprogramming systems [GAVE73,GELE75]. Kobayashi has applied diffusion approximations to general queueing networks [KOBA74a , KOBA74b , KOBA78]. Reiser and Kobayashi [REIS74] report on the accuracy of the diffusion approximation. They found the approximation to yield more accurate results than simply replacing a server originally having a general service time distribution with a server having an exponential service time distribution.

Courtois has studied the analysis of queueing network models through a different approach, decomposition [COUR71 , COUR75 , COUR77]. An example of this approach consists of analyzing a subsystem in isolation and then replacing the entire subsystem in the original model by a single composite server which appears to the rest of the system to behave the same as the original subsystem. Courtois reported that such an approximation is accurate if the rate of interaction within the subsystem is substantially higher than the rate of interactions between the subsystem and the rest of the system.

Another application of decomposition is the Chandy, Herzog, and Woo (CHW) theorem [CHAN75a]. (This is also frequently referred to as Norton's theorem for queueing networks.) Given a network with M stations, the CHW theorem states that it is possible to replace M-1 of the stations with a single composite service center having load dependent service rates. The composite station service rate is chosen so that the performance measures (throughputs, utilizations, mean queue lengths, etc.) at the M-th station remain unchanged from what they were in the original model.

If all stations in the original network have exponential service time distributions (or are operationally homogeneous), then the CHW decomposition is exact [CHAN75a]. In other words, the subsystem replacement has no effect at all on the queue size distributions of any station outside the subsystem. This, of course, is precisely the case in which approximation techniques are unnecessary.

Several papers have addressed the problem of determining the characteristics of the composite service center for networks with general service time distributions [CHAN75b,SAUE75,MARI79]. Sevcik et al. have described several new more accurate approximations to estimate the coefficient of variation at the composite service center [SEVC77]. They emphasize that the departure process of the composite service center must match that of the aggregated subnetwork.

Another technique, an attractive alternative to the CHW theorem, is the extended product form method for solving queueing networks containing FCFS stations with general service time distributions [SHUM77]. Shum and Buzen hypothesize that these queueing networks do, in fact, possess product form solutions. Shum further postulates that the solution of a M/G/1 queueing system with finite waiting room can replace the stations with general service time distributions. The technique employs an iterative search algorithm to determine the arrival rates to the various stations in the network. The assumption of a product form solution implies that efficient computational procedures can be applied to speed up the search. Unfortunately, the search algorithm is not guaranteed to converge. Recently, Balbo has developed a new more efficient search algorithm which has converged quickly in all cases tested [BALB79a].

Balbo presents a thorough study of approximation methods, including numerical methods, in [BALB79b]. Additional information on numerical techniques can be found in [STEW77] and its references.

1.5 CONCLUSION

Our introduction to queueing network models is necessarily brief. An excellent source for additional information on this topic is the September 1978 issue of Computing Surveys [GRAH78].

Although queueing network models cannot deal with some practical cases -- FCFS stations with different service for different job classes, priorities, or blocking of stations -- their computational algorithms make them very attractive as the basis of approximations. Though often more accurate, solutions of global balance equations or diffusion equations are quite tedious. The technique of decomposition allows the analyst to construct approximate models of otherwise intractable real systems by defining stations which are "flow-equivalent" to subsystems of the real system. Once a product-form approximation has been found, the highly efficient computational procedures -- which are the subject of Chapters 2 and 4 of this book -- can be used to calculate (approximate) answers quickly.

CHAPTER 2

COMPUTATIONAL ASPECTS OF CLOSED QUEUEING NETWORKS WITH MULTIPLE JOB CLASSES: THE NORMALIZATION CONSTANT APPROACH

2.1 INTRODUCTION

In this chapter we will consider the class of BCMP closed queueing networks [BASK75]. Informally, these networks permit service centers to have one of four different queueing disciplines and jobs to be grouped into distinct classes. The jobs may change class while moving from one station to another in the network.

Our objective is to present clear, concise, and computationally efficient algorithms for the evaluation of the normalizing constant, marginal queue length distribution, throughput, utilization, mean queue length, and mean waiting time for each station in the network by job class.

The next section reviews the algorithms developed by Buzen for computing performance measures of single job class closed queueing networks [BUZE71b,BUZE73]. In the process of this review, we will describe several important simplifications to these algorithms. Section 2.3 contains an example demonstrating that these procedures can be further simplified for a specific class of networks, those with load independent parameters. Section 2.4 then extends the methods developed in section 2.2 to multiple job class networks (without job class switching). The results of this section are natural generalizations of the single class algorithms developed in section 2.2. Finally, in section 2.5 we extend our analysis techniques to handle networks in which jobs may switch class while traversing the network.

2.2 SINGLE JOB CLASS CLOSED QUEUEING NETWORK MODELS

Before considering multiple job class queueing networks, we will review Buzen's computational algorithms for closed single class queueing networks [BUZE71b,BUZE73]. The review is presented for two reasons. First, an understanding of the single job class computational algorithms is essential for comprehending the more complex multiple job class computational algorithms. Second, the review allows us to present several extensions to Buzen's computational algorithms. The importance of these extensions will become more apparent in the case of multiple job class networks.

The queueing networks being considered contain a finite number M of stations. Each station has an associated queue on which jobs may wait prior to receiving service. A job is present at station i if it is waiting for or receiving service there. We let n_i denote the number of jobs present at station i and $N=n_1+\cdots+n_i+\cdots+n_M$ denote the total in the closed network. A <u>state</u> of the network represents an opportionment of jobs among stations. A <u>feasible state</u> can be described by a vector

$$\underline{n} = (n_1, n_2, \ldots, n_i, \ldots, n_M)$$

in which

$$n_i \geqq 0,\ i=1,2,\ldots,M \quad \text{and} \quad \sum_{i=1}^{M} n_i = N.$$

The fraction of departures from station i that proceed next to station j is termed the transition frequency and is denoted p_{ij} $(i,j=1,2,\ldots,M)$.

Each station i is characterized by a service rate μ_i and a queueing discipline. The service rate is expressed as the number of jobs departing station i per unit of busy time of the station. This service rate may be load dependent, i.e., it may depend upon the length of the associated queue. Then μ_i is a function of the queue length and will be denoted by $\mu_i(k)$. Each station contains one or more servers; for example, there may be multiple CPU's. A multiple server station is load dependent since the rate at which the station completes work increases as more servers in the station become busy. For a load independent station i, the mean service time, s_i, is the reciprocal of the load independent service rate, i.e. $s_i=1/\mu_i$. The queueing discipline determines the order in which jobs at the station receive service. The queueing disciplines for which mathematical results exist are: FCFS, PS, IS, and LCFSPR (see Chapter 1). To avoid unnecessay complications we only treat the analysis of FCFS stations in this section. The other queueing disciplines will be covered in section 2.4.

The assumptions of the closed networks under consideration are summarized below:

> M stations
>
> All stations have FCFS queueing disciplines
>
> N jobs
>
> All service time distributions are exponential
>
> p_{ij} denotes the fraction of departures from station i that proceed next to station j
>
> Service rate at each station may be a function of the station's queue length

The equilibrium distribution of jobs in the network is given by [JACK63,GORD67a,BASK75]:

$$P(\underline{n}) = P(n_1, n_2, \ldots, n_i, \ldots, n_M) = \frac{1}{G(N)} \prod_{i=1}^{M} f_i(n_i)$$

where

$$f_i(n_i) = \begin{cases} \dfrac{e_i^{n_i}}{\prod_{k=1}^{n_i} \mu_i(k)} & \text{if station i is load dependent} \\[2ex] (e_i s_i)^{n_i} = Y_i^{n_i} & \text{if station i is load independent} \end{cases}$$

The e_i are the solutions of the M linear equations:

$$e_j = \sum_{i=1}^{M} e_i p_{ij} \qquad j=1,2,\ldots,M$$

which equate the flows into and out of each station. G(N) is a normalizing constant chosen so that all the $P(\underline{n})$ sum to one.

Since each system state corresponds to an M-tuple $(n_1,n_2,\ldots,n_M)$ of non-negative integers that sum to N, let the set of all possible system states for a closed network with N customers and M stations be defined as

$$S(N,M)=\{(n_1,n_2,\ldots,n_i,\ldots,n_M) \mid \sum_{i=1}^{M} n_i = N,\ n_i \geqq 0\ i=1,\ldots,M\}.$$

The fact that the sum of all the state probabilities must equal one is expressed as

$$1 = \sum_{\underline{n} \in S(N,M)} P(\underline{n}) = \frac{1}{G(N)} \sum_{\underline{n} \in S(N,M)} \prod_{i=1}^{M} f_i(n_i).$$

Hence, the normalization constant is given by

$$G(N) = \sum_{\underline{n} \in S(N,M)} \prod_{i=1}^{M} f_i(n_i).$$

(The notation for the normalization constant indicates that it is only a function of the number of jobs in the network, N. In fact, it depends on M, p_{ij}, e_i, and $\mu_i(k)$ as well.)

All performance measures of the queueing network are dependent upon the basic distribution $P(\underline{n})$. Therefore, the evaluation of the normalization constant G(N) seems essential (of. Chapter 4, though). Unfortunately, a straightforward evaluation of G(N) must be avoided because the number of feasible states $\underline{n}$ in the summation increases rapidly as a function of N and M. As noted by Denning and Buzen [DENN78], each state $\underline{n}=(n_1,n_2,\ldots,n_M)$ can be encoded as a binary string of N 1s and M-1 0s,

$$1^{n_1}01^{n_2}0\cdots01^{n_M};$$

the number of such strings is the number of permutations of N indistinguishable objects and M-1 indistinguishable objects, namely,

$$\binom{N+M-1}{M-1}.$$

Even for the relatively modest case in which M=8 and N=20, the calculation of G(N) requires the summation of 888030 terms, each of which is the product of 8 factors.

Despite the apparently large number of arithmetic operations required to determine G(N), there exists an efficient algorithm which computes not only G(N) but also G(1), G(2), ... ,G(N-1). The procedure was developed by Buzen and has since become known as Buzen's algorithm [BUZE71b,BUZE73]. The next subsections review this algorithm.

2.2.1 REVIEW OF BUZEN'S ALGORITHM

Buzen described his algorithm in two parts, depending on whether the network has load dependent or load independent stations. Rather than adopting his approach directly, we will show that the load independent results are easily obtainable starting from the load dependent formulas.

To derive the algorithm, it is necessary to first define one auxiliary function. Let

$$g_m(n) = \sum_{\underline{n}\in S(n,m)} \prod_{i=1}^{m} f_i(n_i).$$

Note that G(N) is equal to $g_M(N)$ and, in fact, $g_M(n)=G(n)$ for $n=0,1,\ldots,N$. Thus, $g_M(n)$ is the normalization constant for the closed network with n jobs.

Now for $m>1$,

$$g_m(n) = \sum_{\underline{n}\in S(n,m)} \prod_{i=1}^{m} f_i(n_i)$$

$$= \sum_{k=0}^{n} \Big[\sum_{\substack{\underline{n}\in S(n,m)\\ n_m=k}} \prod_{i=1}^{m} f_i(n_i)\Big]$$

$$= \sum_{k=0}^{n} f_m(k)\Big[\sum_{\underline{n}\in S(n-k,m-1)} \prod_{i=1}^{m-1} f_i(n_i)\Big]$$

$$= \sum_{k=0}^{n} f_m(k)\ g_{m-1}(n-k) \qquad (2.1)$$

This expression is used when the service rate of the mth station depends on the load. If the mth station in the network is load independent, then equation 2.1 can be simplified. In this case,

$$f_m(k) = (e_m s_m)^k = Y_m^k = Y_m\ f_m(k-1)$$

since the service rate of the station no longer depends on the load. Substituting this $f_m(k)$ into equation 2.1 and simplifying yields

$$g_m(n) = g_{m-1}(n) + Y_m\ g_m(n-1) \qquad (2.2)$$

when station m is load independent. Note that both expressions for $g_m(n)$ are recurrence relations; i.e. $g_{m-1}(n)$ is required before $g_m(n)$ can be computed. The recurrence relation is properly initialized by noting

$$g_1(n) = \sum_{\underline{n} \in S(n,1)} \prod_{i=1}^{1} f_i(n_i)$$

$$= \sum_{\underline{n} \in \{(n)\}} f_1(n)$$

$$= f_1(n) \qquad \text{for } n=0,1,2,\ldots N.$$

Buzen observes this initial condition [BUZE73] but then does not use it when describing the implementation of the algorithm. Instead he uses $g_0(0)=1$ and $g_0(n)=0$ for $n=1,2,\ldots,N$ to generate $g_1(n)$ using either equation 2.1 or

2.2. This is wasteful especially if station 1 is load dependent.

Figures 2-1 and 2-2 depict the operation of Buzen's algorithm schematically [BUZE73,DENN78]. Interior elements in the two dimensional array are filled in column by column. If station m is load dependent then column m is computed using equation 2.1; otherwise equation 2.2 is employed.

Figure 2-1 suggests an array of dimension (N+1)xM is required. Since the algorithm fills in the array one column at a time, only a vector of size N+1 is needed. This is true even when the network contains load dependent stations. Buzen originally discussed the need for two vectors of size N+1 in this case [BUZE73]. Assuming station m is load dependent, he reasoned that it is necessary to save the entire m-1st column until the last entry in the mth column is computed. But observe that if the elements in the mth column are computed starting with the last element $g_m(N)$ and proceeding to the first element $g_m(1)$, then only N+1 locations are required (cf. [SHUM76]). Note that $g_m(N)$ requires all values in the m-1st column. We can then overwrite $g_{m-1}(N)$ (with $g_m(N)$) because it will no longer be needed in the computation of $g_m(N-1)$, $g_m(N-2)$,...,$g_m(1)$. Likewise, $g_{m-1}(N-1)$ can be overwritten by $g_m(N-1)$.

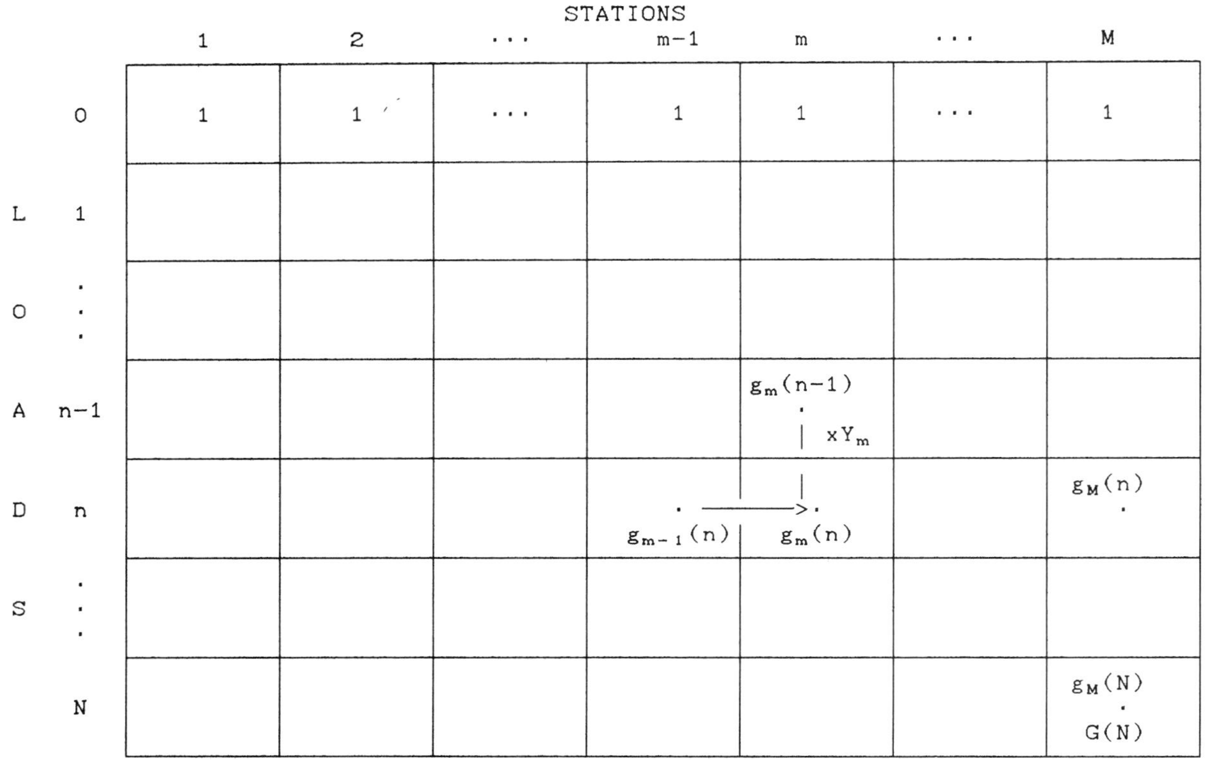

Figure 2-1:

Algorithm for computing normalization constant of a single class closed queueing network if mth station is load independent

STATIONS

LOADS		1	...	m−1	m	...	M
	0	1	...	$g_{m-1}(0) \cdot f_m(n)$ — +			1
L	1			$g_{m-1}(1) \cdot f_m(n-1)$ — +			
O	⋮			⋮			
A	n−1			$g_{m-1}(n-1) \cdot f_m(1)$ — +			
D	n			$g_{m-1}(n) \cdot f_m(0)$ — +→	$\cdot\ g_m(n)$		$g_M(n)$.
S	⋮						
	N						$g_M(N)$. $G(N)$

<u>Figure 2-2</u>:

Algorithm for computing normalization constant of a single class closed queueing network if mth station is load dependent

Table 2-1 provides an implementation of the calculation of the normalization constant for a network with N jobs and M stations. We have used the fact that $f_i(k)$ can be expressed as

$$f_i(k) = \begin{cases} \dfrac{e_i}{\mu_i(k)} f_i(k-1) & \text{if station i is load dependent} \\ \\ Y_i \; f_i(k-1) & \text{if station i is load independent} \end{cases}$$

and that $f_i(0)=1.0$ for $i=1,2,...,M$.

An important simplification to the load dependent formula 2.1 occurs when the mth station contains multiple identical servers. If the mth contains a_m identical servers, then the station's service rate is

$$\mu_m(k) = \begin{cases} k \; \mu_m & \text{if } 0 \leqq k \leqq a_m \\ \\ a_m \; \mu_m & \text{if } k > a_m \end{cases}$$

where μ_m is the service rate of the station with one job. In this case $f_m(k)$ becomes:

$$f_m(k) = \begin{cases} \dfrac{Y_m^{\,k}}{k!} & \text{if } 0 \leqq k \leqq a_m \\ \\ \dfrac{Y_m^{\,k}}{a_m! \, a_m^{\,k-a_m}} & \text{if } k \geqq a_m \end{cases}$$

which can be written as

Table 2-1:
Efficient implementation of Buzen's algorithm

```
(*Program segment to compute normalization constant
 for a network with M stations and N customers.
 We assume:
    STATION[I] = 'LD' for load dependent station
               = 'LI' for load independent station
    MU[I,J]    = service rate of station I with
                 a load of J jobs
    E[I]       = solution to the M linear equations
    S[I]       = mean service time of load
                 independent station I*)

G[0]:=1.0;                         (*gm(0)=1.0 for m=1,...,M*)

if STATION[1] = 'LI' then          (*use load independent*)
begin                              (*initialization of g1(n)*)
  Y:=E[1]*S[1];
  for N1:=1 to N do G[N1]:=G[N1-1]*Y
end
else                               (*use load dependent*)
begin                              (*initialization of g1(n)*)
  Y:=E[1];
  for N1:=1 to N do G[N1]:=G[N1-1]*Y/MU[1,N1]
end;

for M1:=2 to M do                  (*for the other stations*)
begin
  if STATION[M1] = 'LI' then       (*use load independent*)
  begin                            (*algorithm: eq. 2.2*)
     Y:=E[M1]*S[M1];
     for N1:=1 to N do G[N1]:=G[N1]+Y*G[N1-1]
  end
  else                             (*use load dependent*)
  begin                            (*algorithm: eq. 2.1*)
     FM[0]:=1.0; Y:=E[M1];
     for N1:=1 to N do FM[N1]:=FM[N1-1]*Y/MU[M1,N1];
     for N1:=N downto 1 do
     begin
        SUM:=G[N1];
        for K:=1 to N1 do SUM:=SUM+FM[K]*G[N1-K];
        G[N1]:=SUM
     end
  end
end;
```

$$f_m(k) = Y_m \; d_m(k) \; f'_m(k-1) \quad \text{for } k \geqq 1$$

where

$$d_m(k) = \begin{cases} \dfrac{1}{k!} & \text{if } 0 \leqq k \leqq a_m \\[2ex] \dfrac{1}{a_m! \, a_m^{\,k-a_m}} & \text{if } k \geqq a_m+1 \end{cases}$$

and

$$f'_m(k) = Y_m^{\,k} \; .$$

Hence, if station m contains a_m identical servers, then

$$g_m(n) = \sum_{k=0}^{n} f_m(k) \; g_{m-1}(n-k)$$

$$= g_{m-1}(n) + Y_m \sum_{k=1}^{n} d_m(k) \; f'_m(k-1) \; g_{m-1}(n-k)$$

$$= g_{m-1}(n) + Y_m \sum_{k=0}^{n-1} d_m(k+1) \; f'_m(k) \; g_{m-1}(n-1-k)$$

$$= g_{m-1}(n) + Y_m [\sum_{k=0}^{a_m-1} \frac{1}{k+1} f_m(k) \; g_{m-1}(n-1-k)$$

$$+ \sum_{k=a_m}^{n-1} \frac{1}{a_m} f_m(k) \; g_{m-1}(n-1-k) \;]$$

Observe that the last summation may be written as

$$\frac{1}{a_m} g_m(n-1) = \sum_{k=0}^{a_m-1} \frac{1}{a_m} f_m(k) g_{m-1}(n-1-k)$$

$$+ \sum_{k=a_m}^{n-1} \frac{1}{a_m} f_m(k) g_{m-1}(n-1-k)$$

Hence,

$$g_m(n) = g_{m-1}(n)$$

$$+ Y_m \left[\frac{1}{a_m} g_m(n-1) + \sum_{k=0}^{a_m-2} \left(\frac{1}{k+1} - \frac{1}{a_m} \right) f_m(k) g_{m-1}(n-1-k) \right]$$

$$= g_{m-1}(n) + \frac{Y_m}{a_m} [g_m(n-1) + \sum_{k=0}^{a_m-2} \left(\frac{a_m-k-1}{k+1} \right) f_m(k) g_{m-1}(n-1-k)] \quad (2.3)$$

This expression reduces to equation 2.2 when a_m=1. If station m contains two servers (a_m=2), then

$$g_m(n) = g_{m-1}(n) + \frac{Y_m}{2} \left[g_m(n-1) + g_{m-1}(n-1) \right] .$$

When the mth station contains multiple identical servers, equation 2.3 results in considerable computational savings over equation 2.1. In the evaluation of $g_m(n)$, equation 2.1 is employed until $n=a_m-2$. From then on ($n>a_m-2$), equation 2.3 should be used since the summation in this equation is bounded by a_m-2 instead of n.

Up to this point we have reviewed the method of computing the normalization constant of a single class closed queueing network. The computation is performed by an iterative procedure which fills in values in a vector of size N+1.

One iteration is performed for each of the M stations in the network. On the first iteration, the vector, which we have denoted by G, is initialized with the values of the function corresponding to the first station in the network, i.e., $G(n)=f_1(n)$ for $n=0,1,\ldots,N$. Each succeeding iteration computes a new set of values for the vector G from those values stored in G by the preceding iteration. The algorithm used on the mth iteration depends on the characteristics of the mth station. In particular, the algorithm corresponding to equation 2.1, 2.2, or 2.3 is used whenever the mth station is load dependent, load independent, or load dependent with multiple identical servers, respectively. After the Mth iteration, G(N) contains the normalization constant for the network with M stations and N jobs. In fact, as noted previously, all elements of the vector G correspond to normalization constants.

We caution the reader that these normalization constants are <u>not</u> unique since they depend on the particular choice of e_j used in the solution of the job flow balance equations:

$$e_j = \sum_{i=1}^{M} e_i \, p_{ij} \qquad j=1,2,\ldots,M$$

The solution to this set of M linear equations can only be determined up to a multiplicative constant. Hence, if $\underline{e} = (e_1, e_2, \ldots, e_M)$ is a solution to the system, then so is $z\underline{e}$ =

$(ze_1, ze_2, \ldots, ze_M)$ for $z \neq 0$. A particular solution of these equations can be obtained by arbitrarily assigning a value of one to e_1.

2.2.2 MARGINAL QUEUE LENGTH DISTRIBUTION

The marginal probability of finding exactly n jobs at station i is given by

$$p_i(n,N) = \sum_{\substack{\underline{n} \in S(N,M) \\ n_i = n}} P(n_1, n_2, \ldots, n_{i-1}, n, n_{i+1}, \ldots, n_M)$$

for $n=0,1,\ldots,N$. The above expression can be simplified:

$$p_i(n,N) = \frac{1}{G(N)} \sum_{\substack{\underline{n} \in S(N,M) \\ n_i = n}} \prod_{j=1}^{M} f_j(n_j)$$

$$= \frac{f_i(n)}{G(N)} \sum_{\substack{\underline{n} \in S(N,M) \\ n_i = n}} \prod_{\substack{j=1 \\ j \neq i}}^{M} f_j(n_j)$$

To derive an algorithm for computing this marginal distribution requires the definition of another auxiliary function. For a network with N jobs, let

$$g_m^i(n) = \sum_{\substack{\underline{n} \in S(N,m) \\ n_i = N-n}} \prod_{\substack{j=1 \\ j \neq i}}^{m} f_j(n_j)$$

This term can be viewed as a normalization constant of a network related to the original network but with station i removed and with only n jobs. Note carefully that:

$$g_M^M(n) = g_{M-1}(n) \quad \text{for } n=0,1,\ldots,N$$

(See Appendix A for the proof). This property of the auxiliary function will be exploited later to simplify the computation of performance measures for the Mth (or last) station in a network. With this new auxiliary function definition, the marginal distribution becomes

$$p_i(n,N) = \frac{f_i(n)}{G(N)} g_M^i(N-n) \quad \text{for } n=0,1,\ldots,N$$

We are now faced with the problem of finding an algorithm to evaluate this new auxiliary function.

Fortunately, an algorithm similar to the one previously described can be derived. The derivation of this algorithm uses the fact that the marginal probabilities must sum to one. For a closed network with N jobs and M stations,

$$1 = \sum_{k=0}^{N} p_i(k,N) \quad \text{for any } i,\ 1 \leqq i \leqq M$$

$$= \sum_{k=0}^{N} \frac{f_i(k)}{G(N)} g_M^i(N-k)$$

Whence,

$$G(N) = \sum_{k=0}^{N} f_i(k)\, g_M^i(N-k)$$

and

$$g_M^i(N) = G(N) - \sum_{k=1}^{N} f_i(k)\, g_M^i(N-k) \tag{2.4}$$

Values for this auxiliary function are computed iteratively

with the initial condition

$$g_M^i(0) = G(0) = 1.0.$$

Equation 2.4 represents the iterative procedure employed when station i is load dependent. As before, simplifications result when station i is load independent or contains multiple identical servers. Rather than explicitly deriving these expressions, we simply list them here:

if station i is load independent:

$$g_M^i(n) = G(n) - Y_i\, G(n-1) \tag{2.5}$$

if station i contains a_i identical servers:

$$g_M^i(n)=G(n) - \frac{Y_i}{a_i}[G(n-1) + \sum_{k=0}^{a_i-2} \left(\frac{a_i-k-1}{k+1}\right) f_i(k) g_M^i(n-1-k)] \tag{2.6}$$

if station i contains two servers (a_i=2):

$$g_M^i(n) = G(n) - \frac{Y_i}{2}\,[G(n-1) + g_M^i(n-1)]$$

It thus becomes evident that the marginal probability of finding exactly n jobs at station i can be computed as

$$p_i(n,N) = \begin{cases} \dfrac{Y_i^n}{G(N)}\,[G(N-n) - Y_i G(N-1-n)] & \text{if station i is load independent} \\[2ex] \dfrac{f_i(n)}{G(N)}\, g_M^i(N-n) & \text{if station i is load dependent} \end{cases}$$

where it is assumed that G(n) is defined to be zero if n<0.

The load independent expression corresponds to Buzen's formula [BUZE71b, BUZE73]. Buzen determined the load dependent marginal probability only for the Mth station in the network. He suggested computing the marginal probability for any other load dependent station by permuting the order in which the stations are numbered so that the station of interest is designated as the Mth station. Unfortunately, his algorithm must then be reapplied to this permuted sequence of stations. Our method avoids this (possibly expensive) reindexing operation entirely. Avoiding reindexing and recomputing steps of the iterative algorithm is crucial to the efficiency of the multiple job class computational algorithms.

The next sections deal with computing the performance measures of a closed queueing network. We will derive expressions for throughput, utilization, mean queue length, mean waiting time, and the mean number of active servers at each station. These performance measures are simple functions of the normalization constant(s).

2.2.3 THROUGHPUT

The throughput of a station measures the rate at which jobs leave that station. Let $X_i(N)$ denote the throughput of station i in a network with N jobs. Then from the definition of throughput:

$$X_i(N) = \sum_{n=1}^{N} p_i(n,N)\ \mu_i(n)$$

$$= \sum_{n=1}^{N} \frac{f_i(n)}{G(N)}\ g_M^i(N-n)\ \mu_i(n)$$

$$= \sum_{n=1}^{N} \frac{e_i}{\mu_i(n)}\ \frac{f_i(n-1)}{G(N)}\ g_M^i(N-n)\ \mu_i(n)$$

$$= \frac{e_i}{G(N)} \sum_{n=1}^{N} f_i(n-1)\ g_M^i(N-n)$$

$$= \frac{e_i}{G(N)} \sum_{n=0}^{N-1} f_i(n)\ g_M^i(N-1-n)$$

$$= e_i \frac{G(N-1)}{G(N)}$$

Thus, the calculation of throughput is a simple function of the normalization constant and the job flow equations. This expression works whether the station is load independent or not.

Recall from section 2.2.1 that the e_i could not be determined uniquely, i.e., there are an infinite number of solutions to the job flow balance equations. The above formula shows that for a network with N jobs, when a particular solution e_i is multiplied by $z=G(N-1)/G(N)$, then the actual station throughputs result.

2.2.4 UTILIZATION

Let $U_i(N)$ denote the utilization of station i in a network with N jobs. Buzen proved that the utilization of a load independent station is [BUZE76]:

$$U_i(N) = s_i \cdot X_i(N)$$

Thus, on substituting the above throughput expression and recalling that $Y_i = e_i s_i$, utilization can also be expressed:

$$U_i(N) = Y_i \cdot \frac{G(N-1)}{G(N)}$$

This is Buzen's previously published result [BUZE71b,BUZE73].

The utilization of a load dependent station follows from the fact that utilization is simply the probability that the station is active. Hence,

$$U_i(N) = \sum_{n=1}^{N} p_i(n,N)$$

$$= 1 - p_i(0,N)$$

$$= 1 - \frac{g_M^i(N)}{G(N)}$$

If the Mth station is load dependent, this expression simplifies somewhat. Recall that:

$$g_M^M(n) = g_{M-1}(n) \quad \text{for } n=0,1,\ldots,N \ .$$

Hence, the utilization of the Mth station is

$$U_M(N) = 1 - \frac{g_{M-1}(N)}{G(N)} \ .$$

2.2.5 MEAN QUEUE LENGTH

The mean queue length at station i when there are N jobs in the network is

$$\bar{n}_i(N) = \sum_{n=1}^{N} n \, p_i(n,N) \ .$$

For a load independent service center, Buzen shows that the mean queue length calculation reduces to [BUZE71b,BUZE73]:

$$\bar{n}_i(N) = \sum_{n=1}^{N} Y_i^n \frac{G(N-n)}{G(N)} \ .$$

From this form, it is a routine exercise to derive the following recursive evaluation for the mean queue length:

$$\bar{n}_i(N) = U_i(N) \ (1+\bar{n}_i(N-1))$$

with the initial condition $\bar{n}_i(0)=0$.

For a load dependent station, the mean queue length must be evaluated from its definition. But, if the Mth station is load dependent, the computation of mean queue length can be sped-up by performing it in conjunction with the evaluation of the normalization constant. To see this note:

$$\bar{n}_M(N) = \sum_{n=1}^{N} n \; p_M(n,N)$$

$$= \sum_{n=1}^{N} n \frac{f_M(n)}{G(N)} \; g_M^M(N-n)$$

$$= \frac{1}{G(N)} \sum_{n=1}^{N} n \; f_M(n) \; g_{M-1}(N-n)$$

Then notice the striking similarity between the last summation and the last step in the computation of the normalization constant

$$G(N) = \sum_{n=0}^{N} f_M(n) \; g_{M-1}(N-n) \; .$$

2.2.6 MEAN WAITING TIME

Once the throughput and mean queue length of a station are computed, the mean waiting time at the station follows directly from Little's formula [LITT61]. Little's result states that the average number of jobs (customers) in a "queueing system" is equal to the average arrival rate of jobs to that system times the average time spent in that system [KLEI75]. Since the average arrival rate in a system in equilibrium must equal the average departure rate (or throughput), Little's result can be written:

$$\bar{n} = X\,\bar{w}.$$

If we let the "queueing system" comprise both the queue and the server(s) at station i, then

$$\bar{w}_i(N) = \frac{\bar{n}_i(N)}{X_i(N)} .$$

Thus, our interpretation of mean waiting time at station i includes both time spent waiting in queue <u>and</u> time spent in service.

The mean waiting time can also be written as a recursion if station i is load independent, viz.,

$$\bar{w}_i(N) = s_i\ (1+\bar{n}_i(N-1)) .$$

This expression is derived by substituting the load independent formulas for the mean queue length and throughput in the previous equation. The mean waiting time a job experiences at a load independent station is thus simply its average service time (s_i) plus the mean time spent by the previously queued jobs ($s_i\ \bar{n}_i(N-1)$).

The recursive evaluation of the mean waiting times is the basis of the mean value analysis approach to solving queueing network models. This method demonstrates that for networks with load independent stations, the computation of all the performance measures does <u>not</u> require the evaluation of the normalization constant! Reiser and Lavenberg were the first to prove a portion of this result [REIS78]. Chapter 4 is devoted to a detailed treatment of this technique.

2.3 MULTIPLE JOB CLASS CLOSED QUEUEING NETWORK MODELS

We now demonstrate that the methods developed for analyzing single job class networks can be extended to handle multiple job class networks. Initially we restrict our attention to closed networks which do not allow jobs to change class membership. The latter topic will be treated in a subsequent section.

Again, the queueing networks being considered contain a finite number M of stations. There are four different types of stations depending on the queueing (or service) discipline. The permissible service disciplines are first come first served (FCFS), processor sharing (PS), infinite servers (IS), and last come first served preemptive-resume (LCFS). All jobs have the same exponential service time distribution at a FCFS service center. Each class of job may have a distinct service time distribution at the other types of stations. The only restriction is that the service time distribution must have a rational Laplace transform. A PS station contains a single server which is shared by jobs infinitely fast in round robin fashion. The net effect is that when there are n jobs at the station, each is receiving service at a rate of 1/n sec/sec. For an IS station each job receives its own server immediately. That is, there are always at least as many servers as jobs. Hence, no queueing delays arise at such a station. These stations are typically used to represent user terminals in a time-sharing model.

The model permits the definition of multiple classes of customers whose characteristics (service rates and transition frequencies) may be distinct. Let there be R distinct classes numbered 1,2,...,R.

Each job class has defined for it a separate set of frequencies governing the movement of jobs of that class between stations. More specifically, a job of class r which completes service at station i will next require service from station j with a certain transition frequency denoted by $p_{ij}(r)$.

In a closed queueing network the total number of jobs remains fixed at some number N. If we let N_r denote the number of class r jobs, then

$$\sum_{r=1}^{R} N_r = N.$$

A state, $\underline{\underline{n}}$, in such a model is a feasible distribution of jobs over classes and stations. More formally, the model is said to be in state $\underline{\underline{n}}=(\underline{n}_1,\underline{n}_2,\ldots,\underline{n}_i,\ldots,\underline{n}_M)$, where $\underline{n}_i=(n_{i1},n_{i2},\ldots,n_{ir},\ldots,n_{iR})$, when there are n_{ir} jobs of class r at service center i. The n_{ir} must satisfy four constraints for the state to be feasible (or meaningful). First, the number of jobs of each class at each station must be nonnegative (i.e., $n_{ir}\geqq 0$, for $i=1,2,\ldots,M$ and $r=1,2,\ldots,R$). Second, $n_{ir}>0$ only if it is possible (under the transition rules of the network) for jobs of class r to be at station i. Third, the total number of jobs in a state must be equal to N, i.e.,

$$\sum_{r=1}^{R}\sum_{i=1}^{M} n_{ir} = N.$$

Fourth, the sum of the number of class r jobs at all the stations must be N_r, i.e.,

$$\sum_{i=1}^{M} n_{ir} = N_r \quad \text{for } r=1,2,\ldots,R.$$

Now we can explicitly define the set of <u>all</u> feasible states by

$$S(N_1,N_2,\ldots,N_R,M)=\{(\underline{n}_1,\underline{n}_2,\ldots,\underline{n}_i,\ldots,\underline{n}_M)\,|\,0\leqq n_{ir}\leqq N_r \text{ for } 1\leqq i\leqq M,$$

$$1\leqq r\leqq R \text{ and } \sum_{i=1}^{M} n_{ir} = N_r \text{ for } 1\leqq r\leqq R\}.$$

Below we summarize the assumptions of the closed networks under consideration.

> M stations
>
> Four different queueing disciplines
>
> FCFS (first come first served)
> PS (processor sharing)
> IS (infinite servers)
> LCFS (last come first served)
>
> R distinct job classes
>
> General service time distributions at PS, IS, and LCFS stations. All classes of jobs must have the same exponential service time distribution at FCFS stations.
>
> $p_{ij}(r)$ denotes the fraction of class r job departures from station i that proceed next to station j

The equilibrium probability of the state $\underline{\underline{n}}=(\underline{n}_1,\underline{n}_2,\ldots,\underline{n}_i,\ldots,\underline{n}_M)$ is :

$$P(\underline{\underline{n}})=P(\underline{n}_1,\underline{n}_2,\ldots,\underline{n}_i,\ldots,\underline{n}_M)=\frac{1}{G(N_1,N_2,\ldots,N_R)}\prod_{i=1}^{M} f_i(\underline{n}_i)$$

where

$$f_i(\underline{n}_i) = \begin{cases} \dfrac{n_i!}{\prod_{k=1}^{n_i} \mu_i(k)} \prod_{r=1}^{R} \dfrac{1}{n_{ir}!} e_{ir}^{\,n_{ir}} & \text{if station i is load dependent and has a FCFS, PS, or LCFS service discipline} \\[2ex] n_i! \prod_{r=1}^{R} \dfrac{1}{n_{ir}!} (e_{ir}s_{ir})^{n_{ir}} & \text{if station i is load independent and has a FCFS, PS, or LCFS service discipline} \\[2ex] \prod_{r=1}^{R} \dfrac{1}{n_{ir}!} (e_{ir}s_{ir})^{n_{ir}} & \text{if station i has an IS service discipline} \end{cases}$$

It is clear that we have augmented our notation to include a job class designation (the subscript r). Thus, s_{ir} denotes the mean service time for a class r job at station i. <u>At a FCFS station though, all jobs must have the same mean service time, i.e., $s_{i1}=s_{i2}=\cdots=s_{iR}$</u>. The e_{ir} are the solutions of the R sets of linear equations:

$$e_{jr} = \sum_{i=1}^{M} e_{ir}\, p_{ij}(r) \quad j=1,2,\ldots,M \text{ and } r=1,2,\ldots,R.$$

Each of the R sets contains M linear equations. These equations balance the flows into and out of each station by each of the R job classes.

Also, n_i now denotes the total number of jobs of any class at service center i. That is,

$$n_i = \sum_{r=1}^{R} n_{ir}\,.$$

$G(N_1,N_2,\ldots,N_R)$ is a normalizing constant chosen so that all the $P(\underline{\underline{n}})$ sum to one. We shall also denote the product $e_{ir} \cdot s_{ir}$ by Y_{ir}.

The above result was proven by Baskett, Chandy, Muntz, and Palacios. It is often referred to as the BCMP theorem. The implications of the theorem should not be overlooked. As stated in [BASK75]:

> "Although we began with almost general service time distributions for PS, IS, and LCFS service centers, only the <u>mean</u> service times appear in $p(\underline{\underline{n}} = (\underline{n}_1, \underline{n}_2, \ldots, \underline{n}_M))$. Thus for the aggregate states and within the bounds of the assumptions of the model, any service time distributions for the different classes of customers yield the same results as exponential service time distributions! It is important to note that while only the means of the service time distribution appear in the results, the effects of the different classes of customers is still present, i.e., the means s_{ir} appear in the solution."

Also, note the "product form" of state probabilities, where the probability of a given state is the product of terms each of which relates to the conditions prevailing at one particular service center.

The next sections extend the results of section 2.2 to multiple job class networks when customers are not permitted to switch class membership.

2.3.1 COMPUTATION OF THE NORMALIZATION CONSTANT

In this section an efficient iterative algorithm is developed for the computation of the normalization constant for a multiple job class closed network.

Again, the sum of all the state probabilities must equal one. Hence,

$$1 = \sum_{\underline{n} \in S(N_1, N_2, \ldots, N_R, M)} P(\underline{n})$$

$$= \frac{1}{G(N_1, N_2, \ldots, N_R)} \sum_{\underline{n} \in S(N_1, N_2, \ldots, N_R, M)} \prod_{i=1}^{M} f_i(\underline{n}_i).$$

In order to reduce the length of many of the following derivations, we introduce a compact vector notation. In particular, let

$$\underline{N}_R \equiv (N_1, N_2, \ldots, N_R)$$

$$\underline{n}_R \equiv (n_1, n_2, \ldots, n_R)$$

$$\underline{k}_R \equiv (k_1, k_2, \ldots, k_R)$$

$$\underline{n}_R = \underline{k}_R \equiv n_1 = k_1,\ n_2 = k_2,\ \ldots,\ n_R = k_R$$

$$\underline{0}_R \equiv (0, 0, \ldots, 0)$$ (i.e. a vector of R zeroes)

$$\underline{1}_r \equiv (0, 0, \ldots, 1, \ldots, 0)$$ (i.e. a 1 in the rth component and R-1 zeroes)

$$\sum_{\underline{k}_R=\underline{0}_R}^{\underline{N}_R} \equiv \sum_{k_R=0}^{N_R} \cdots \sum_{k_2=0}^{N_2} \sum_{k_1=0}^{N_1}$$

$$\delta(\underline{n}_R) \equiv \begin{cases} 0 & \text{if } n_1=n_2=\cdots=n_R=0 \\ 1 & \text{else} \end{cases}$$

Employing this notation, the normalization constant becomes

$$G(\underline{N}_R) = \sum_{\underline{n}\in S(\underline{N}_R,M)} \prod_{i=1}^{M} f_i(\underline{n}_i)$$

With the addition of job classes, we note that the size of the normalization constant "vector" increases. Now rather than a vector (of size N+1 elements), we require an R dimensional array. The structure of the array is given by the PASCAL-like declaration:

G : <u>array</u> [0..N_1,0..N_2,...,0..N_R] <u>of</u> <u>real</u>.

Hence, the array size is

$$\prod_{r=1}^{R} (N_r + 1) \quad \text{elements.}$$

It is not difficult to find a distribution of jobs over classes such that the number of locations required by this array to exceed the capacity of a computer's main memory. This is a genuine problem for networks which have many job classes and jobs. Fortunately, for the types of networks of interest to computer system modellers, only a moderate size array is ever required (typically less than 3000 locations).

Again, an auxiliary function is essential to the derivation of a computational procedure. Let

$$g_m(\underline{n}_R) = \sum_{\underline{\underline{n}} \in S(\underline{n}_R, m)} \prod_{i=1}^{m} f_i(\underline{n}_i)$$

Note that $G(\underline{N}_R)=g_M(\underline{N}_R)$ and, in fact, $G(\underline{n}_R)=g_M(\underline{n}_R)$ for any permissible value of n_r $(1 \leqq r \leqq R)$.

Now for $m>1$,

$$g_m(\underline{n}_R) = \sum_{\underline{\underline{n}} \in S(\underline{n}_R, m)} \prod_{i=1}^{m} f_i(\underline{n}_i)$$

$$= \sum_{\underline{k}_R=\underline{0}_R}^{\underline{n}_R} \Big[\sum_{\substack{\underline{\underline{n}} \in S(\underline{n}_R, m) \\ \underline{n}_m=\underline{k}_R}} \prod_{i=1}^{m} f_i(\underline{n}_i) \Big]$$

$$= \sum_{\underline{k}_R=\underline{0}_R}^{\underline{n}_R} f_m(\underline{k}_R) \Big[\sum_{\underline{\underline{n}} \in S(\underline{n}_R-\underline{k}_R, m-1)} \prod_{i=1}^{m-1} f_i(\underline{n}_i) \Big]$$

$$= \sum_{\underline{k}_R=\underline{0}_R}^{\underline{n}_R} f_m(\underline{k}_R)\, g_{m-1}(\underline{n}_R-\underline{k}_R)$$

Explicitly,

$$g_m(n_1, n_2, \ldots, n_R) = \sum_{k_R=0}^{n_R} \cdots \sum_{k_2=0}^{n_2} \sum_{k_1=0}^{n_1} f_m(k_1, k_2, \ldots, k_R) \cdot g_{m-1}(n_1-k_1, n_2-k_2, \ldots, n_R-k_R) \qquad (2.7)$$

When the service rate of the mth station is load dependent this expression must be used. Fortunately, simplifications to this result exist when station m is either load independent or contains multiple identical servers.

To prove this, we require the following lemma.

Lemma 1: If station m is load independent and has a FCFS, PS, or LCFS service discipline, then

$$f_m(\underline{n}_R) = \sum_{r=1}^{R} Y_{mr} \, f_m(\underline{n}_R - \underline{1}_r) \ .$$

If any component of the vector $\underline{n}_R - \underline{1}_r$ is negative, then the value of the function is zero.

Proof: See Appendix A.

Then when station m is load independent, and not an IS station,

$$g_m(\underline{n}_R) = \sum_{\underline{k}_R = \underline{0}_R}^{\underline{n}_R} f_m(\underline{k}_R) \, g_{m-1}(\underline{n}_R - \underline{k}_R)$$

$$= g_{m-1}(\underline{n}_R) + \sum_{\underline{k}_R = \underline{0}_R}^{\underline{n}_R} \delta(\underline{k}_R) f_m(\underline{k}_R) g_{m-1}(\underline{n}_R - \underline{k}_R)$$

$$= g_{m-1}(\underline{n}_R) + \sum_{r=1}^{R} Y_{mr} \left[\sum_{\underline{k}_R = \underline{0}_R}^{\underline{n}_R} \delta(\underline{k}_R) f_m(\underline{k}_R - \underline{1}_r) g_{m-1}(\underline{n}_R - \underline{k}_R) \right]$$

$$= g_{m-1}(\underline{n}_R) + \sum_{r=1}^{R} Y_{mr} \left[\sum_{\underline{k}_R = \underline{1}_r}^{\underline{n}_R} f_m(\underline{k}_R - \underline{1}_r) \, g_{m-1}(\underline{n}_R - \underline{k}_R) \right]$$

$$= g_{m-1}(\underline{n}_R) + \sum_{r=1}^{R} Y_{mr} \left[\sum_{\underline{k}_R = \underline{0}_R}^{\underline{n}_R - \underline{1}_r} f_m(\underline{k}_R) \, g_{m-1}(\underline{n}_R - \underline{1}_r - \underline{k}_R) \right]$$

$$= g_{m-1}(\underline{n}_R) + \sum_{r=1}^{R} Y_{mr}\ g_m(\underline{n}_R - \underline{1}_r)$$

Expanding this result,

$$g_m(n_1, n_2, \ldots, n_R) = g_{m-1}(n_1, n_2, \ldots, n_R) + \sum_{r=1}^{R} Y_{mr}\ g_m(n_1, n_2, \ldots, n_r - 1, \ldots, n_R) \qquad (2.8)$$

In the evaluation of equation 2.8 if n_r-1 is less than zero, then $g_m(\cdot,\cdot,\ldots,n_r-1,\ldots,\cdot)$ is defined to be zero. This statement applies to all succeeding expressions as well and will therefore not be repeated each time. Figure 2-3 depicts the operation of this load independent algorithm for a closed network with 2 job classes. Note from equation 2.8 and Figure 2-3 that to fill in the current matrix (g_m) numbers from the previous g matrix (g_{m-1}) are required.

In Appendix A we prove that if the mth station contains a_m multiple identical servers, then

$$g_m(\underline{n}_R) = g_{m-1}(\underline{n}_R) + \frac{1}{a_m} \sum_{r=1}^{R} Y_{mr} \Big[g_m(\underline{n}_R - \underline{1}_r) + \sum_{\substack{\underline{k}_R=\underline{0}_R \\ k_m \leq a_m-2}}^{\underline{n}_R-\underline{1}_r} \left(\frac{a_m - k_m - 1}{k_m + 1} \right) f_m(\underline{k}_R)\ g_{m-1}(\underline{n}_R - \underline{1}_r - \underline{k}_R) \Big] \qquad (2.9)$$

where $k_m = k_1 + k_2 + \cdots + k_R$. The last summation denotes R nested sums. The condition $k_m \leq a_m - 2$ implies that the upper bound of the rth sum ($r=1,2,\ldots,R$) need only range to $\min(n_r, a_m-2)$.

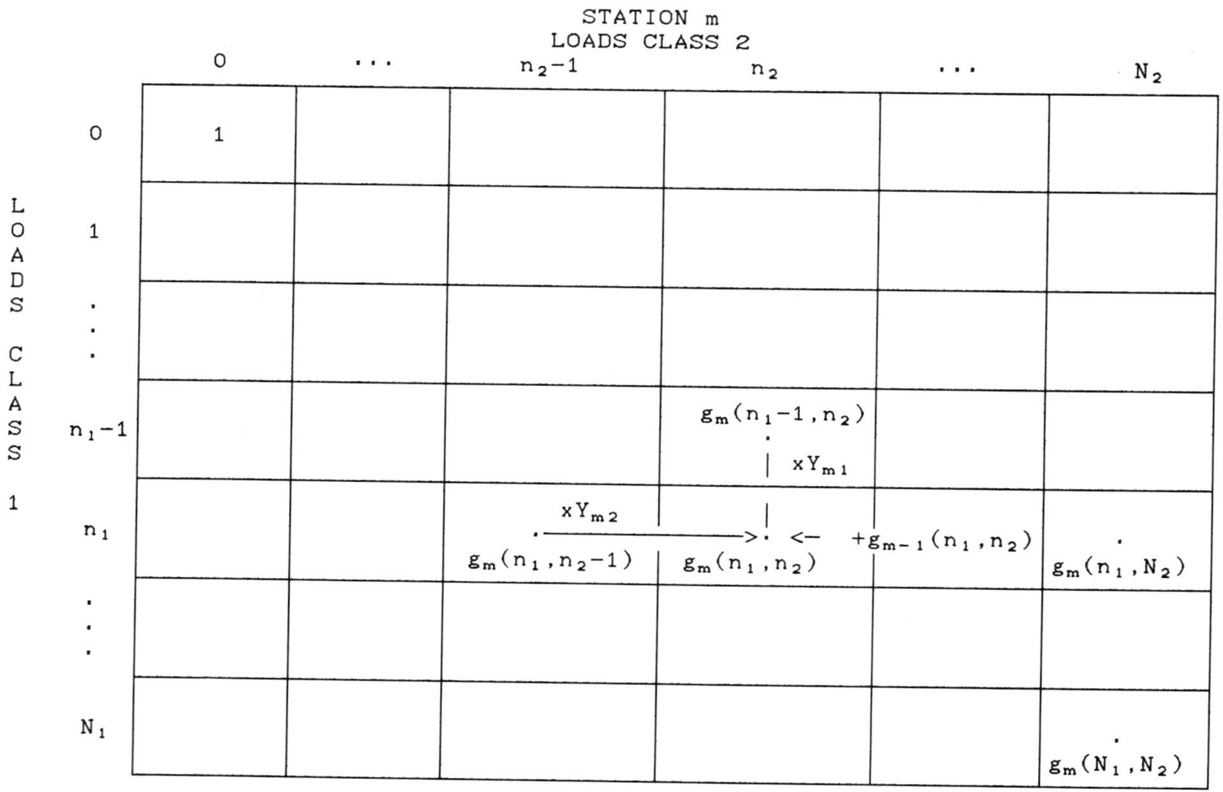

Figure 2-3:
Algorithm for computing normalization constant of a two job class closed queueing network if mth station is load independent

To properly initialize the recurrence relations of equations 2.7, 2.8, and 2.9 we note that

$$g_1(\underline{n}_R) = \sum_{\underline{\underline{n}} \in S(\underline{n}_R, 1)} \prod_{i=1}^{1} f_i(\underline{n}_i)$$

$$= \sum_{\underline{\underline{n}} \in \{(\underline{n}_R)\}} f_1(\underline{n}_R)$$

$$= f_1(\underline{n}_R)$$

or

$$g_1(n_1, n_2, \ldots, n_R) = f_1(n_1, n_2, \ldots, n_R) \text{ for } \begin{cases} n_1 = 0, 1, \ldots, N_1 \\ \cdots\cdots \\ n_R = 0, 1, \ldots, N_R \end{cases}$$

Now consider the case when the mth station of the network is an IS station. Unfortunately, equation 2.7 must be used whenever an IS is not the first station in the network. This can result in a considerable amount of computation. If there is only one IS station in the network, the easiest way to avoid invoking equation 2.7 is to index the network so that the IS station is the first service center. Hence, the above initialization procedure will be employed. But the problem still exists if there are multiple IS stations. In a latter section we will demonstrate that all IS stations can be coalesced into a single IS station; we then let the new IS station be the first service center in the network, thereby eliminating the above computational problem.

2.3.2 MARGINAL QUEUE LENGTH DISTRIBUTION

The marginal probability of finding n_r jobs of class r (r=1,2,...,R) at station i is given by

$$p_i(\underline{n}_R,\underline{N}_R) = \sum_{\substack{\underline{n}\in S(\underline{N}_R,M)\\ \underline{n}_i=\underline{n}_R}} P(\underline{n}_1,\underline{n}_2,\ldots,\underline{n}_{i-1},\underline{n}_R,\underline{n}_{i+1},\ldots,\underline{n}_M)$$

$$= \frac{f_i(\underline{n}_R)}{G(\underline{N}_R)} \sum_{\substack{\underline{n}\in S(\underline{N}_R,M)\\ \underline{n}_i=\underline{n}_R}} \prod_{\substack{j=1\\ j\neq i}}^{M} f_j(\underline{n}_j)$$

Generalizing the auxiliary function definition to multiple job classes yields:

$$g_m^{i}(\underline{n}_R) = \sum_{\substack{\underline{n}\in S(\underline{N}_R,m)\\ \underline{n}_i=\underline{N}_R-\underline{n}_R}} \prod_{\substack{j=1\\ j\neq i}}^{m} f_j(\underline{n}_j)$$

With this auxiliary function definition, the marginal distribution becomes

$$p_i(\underline{n}_R,\underline{N}_R) = \frac{f_i(\underline{n}_R)}{G(\underline{N}_R)}\, g_M^{i}(\underline{N}_R-\underline{n}_R) \quad \text{for } \underline{n}_R=\underline{0}_R,\ldots,\underline{N}_R$$

Again, the derivation of an iterative algorithm to compute the auxiliary function relies on the marginal probabilities summing to one. For a closed network with $N_1,N_2,\ldots,N_R$ jobs and M stations,

$$1 = \sum_{\underline{k}_R=\underline{0}_R}^{\underline{N}_R} p_i(\underline{k}_R,\underline{N}_R) \quad \text{for any i, } 1\leq i\leq M$$

Hence,

$$G(\underline{N}_R) = \sum_{\underline{k}_R=\underline{0}_R}^{\underline{N}_R} f_i(\underline{k}_R)\, g_M^i(\underline{N}_R-\underline{k}_R)$$

and

$$g_M^i(\underline{N}_R) = G(\underline{N}_R) - \sum_{\underline{k}_R=\underline{0}_R}^{\underline{N}_R} \delta(\underline{k}_R) f_i(\underline{k}_R) g_M^i(\underline{N}_R-\underline{k}_R) \qquad (2.10)$$

Values for this auxiliary function are computed iteratively starting with the initial condition

$$g_M^i(\underline{0}_R) = G(\underline{0}_R) = 1.0$$

We note in passing that the $\delta(\underline{k}_R)$ appearing in equation 2.10 is superfluous if we make the following observation: to compute any $g_M^i(\underline{N}_R)$ the only time $\delta(\underline{k}_R)$ is zero is when $k_1=k_2=\cdots=k_R$, i.e., when we have $g_M^i(\underline{N}_R)$ on the right hand side of the equation. Therefore, we may simply initialize $g_M^i(\underline{N}_R)$ to zero, before applying eq. 2.10.

As usual, simplifications result when station i is load independent or contains multiple identical servers. In particular,

if station i is load independent:

$$g_M^i(\underline{n}_R) = G(\underline{n}_R) - \sum_{r=1}^{R} Y_{ir}\, G(\underline{n}_R - \underline{1}_r) \qquad (2.11)$$

if station i contains a_i identical servers:

$$g_M^i(\underline{n}_R) = G(\underline{n}_R) - \frac{1}{a_i} \sum_{r=1}^{R} Y_{ir} \Big[G(\underline{n}_R - \underline{1}_r) + \sum_{\substack{\underline{k}_R=\underline{0}_R \\ k_i \leq a_i - 2}}^{\underline{n}_R - \underline{1}_r} \left(\frac{a_i - k_i - 1}{k_i + 1} \right) f_i(\underline{k}_R)\, g_M^i(\underline{n}_R - \underline{1}_r - \underline{k}_R) \Big] \tag{2.12}$$

It is thus apparent that the marginal probability of finding exactly $\underline{n}_R$ jobs at station i can be computed as

$$p_i(\underline{n}_R, \underline{N}_R) = \begin{cases} \dfrac{f_i(\underline{n}_R)}{G(\underline{N}_R)} \Big[G(\underline{N}_R - \underline{n}_R) - \displaystyle\sum_{r=1}^{R} Y_{ir} G(\underline{N}_R - \underline{1}_r - \underline{n}_R) \Big] & \text{if station i is load independent} \\[2ex] \dfrac{f_i(\underline{n}_R)}{G(\underline{N}_R)}\, g_M^i(\underline{N}_R - \underline{n}_R) & \text{if station i is load dependent} \end{cases}$$

The next subsections deal with computing performance measures of multiple job class closed queueing networks.

2.3.3 THROUGHPUT

Let $X_{ir}(\underline{N}_R)$ denote the throughput of class r jobs at station i in a network with $\underline{N}_R$ jobs. The marginal probabilities which we have computed yield the probability of the configuration $\underline{n}_R=(n_1, n_2, \ldots, n_R)$ which is independent of the actual ordering of jobs at the station. To compute the throughput of class r jobs at station i, we need to know the probability of a class r job being in service. This is the reason the term n_{ir}/n_i appears in the following formulas. The next lemma will be used in the throughput derivation.

Lemma 2: For any FCFS, PS, LCFS, or IS station i ($1 \leq i \leq M$) and for any job class r ($1 \leq r \leq R$), if $n_{ir} \neq 0$, then

$$f_i(\underline{n}_i) = \begin{cases} e_{ir} \dfrac{n_i \, f_i(\underline{n}_i - \underline{1}_r)}{n_{ir}\mu_i(n_i)} & \text{if station i has a FCFS, PS, or LCFS service discipline} \\ e_{ir} \dfrac{s_{ir}}{n_i} f_i(\underline{n}_i - \underline{1}_r) & \text{if station i has an IS service discipline} \end{cases}$$

Proof: See Appendix A.

The throughput of class r jobs at station i is

$$X_{ir}(\underline{N}_R) = \sum_{\underline{n}_R = \underline{1}_r}^{\underline{N}_R} p_i(\underline{n}_R, \underline{N}_R) \frac{n_{ir}}{n_i} \mu_i(n_i)$$

$$\vdots$$

$$= e_{ir} \frac{G(\underline{N}_R - \underline{1}_r)}{G(\underline{N}_R)}$$

$$= e_{ir} \frac{G(N_1, N_2, \ldots, N_r - 1, \ldots, N_R)}{G(N_1, N_2, \ldots, N_R)}$$

The proof is straightforward and can be found in Appendix A.

This result applies for any of the four types of queueing disciplines. It also applies for stations with load dependent or load independent service rates!

2.3.4 UTILIZATION

Let $U_{ir}(\underline{N}_R)$ denote the utilization of station i by class r jobs in a network with $\underline{N}_R$ jobs. (This measure does not apply for an IS station). Generalizing Buzen's proof of the utilization of a single class load independent station, we deduce

$$U_{ir}(\underline{N}_R) = s_{ir} \cdot X_{ir}(\underline{N}_R) .$$

On substituting the above expression for throughput and recalling that $Y_{ir} = e_{ir}s_{ir}$, we see that utilization could also be computed:

$$U_{ir}(\underline{N}_R) = Y_{ir} \frac{G(\underline{N}_R - \underline{1}_r)}{G(\underline{N}_R)} .$$

For a load dependent station, utilization must be calculated from its definition, i.e.,

$$U_{ir}(\underline{N}_R) = \sum_{\underline{n}_R = \underline{1}_r}^{\underline{N}_R} p_i(\underline{n}_R, \underline{N}_R) \frac{n_{ir}}{n_i}$$

As Wong explains [WONG75], the term n_{ir}/n_i is the fraction of the server capacity received by class r jobs when station i has a PS service discipline; the quotient is also the fraction of time that a class r job is getting full capacity of the server when station i has a FCFS or LCFS service discipline.

2.3.5 MEAN QUEUE LENGTH

Let $\bar{n}_{ir}(\underline{N}_R)$ denote the average number of class r jobs at service center i in a network with $\underline{N}_R$ jobs. For a station with a FCFS, PS, or LCFS queueing discipline, the definition of mean queue length by job class is:

$$\bar{n}_{ir}(\underline{N}_R) = \sum_{\underline{n}_R=\underline{1}_r}^{\underline{N}_R} p_i(\underline{n}_R,\underline{N}_R)\ n_r$$

In Appendix A we prove that if station i is load independent, then

$$\bar{n}_{ir}(\underline{N}_R) = U_{ir}(\underline{N}_R)\ (1 + \bar{n}_i(\underline{N}_R-\underline{1}_r))$$

where $\bar{n}_i(\underline{N}_R)$ denotes the <u>total</u> mean queue length at station i when there are $\underline{N}_R$ jobs in the network. This total mean queue length may be computed from

$$\bar{n}_i(\underline{N}_R) = \sum_{r=1}^{R} U_{ir}(\underline{N}_R)\ (1 + \bar{n}_i(\underline{N}_R-\underline{1}_r))$$

with the initial condition $\bar{n}_i(\underline{0}_R)=0.0$.

As before, if the Mth station is load dependent, the computation of the mean queue length can be performed in conjunction with the evaluation of the normalization constant, since

$$\bar{n}_{Mr}(\underline{N}_R) = \sum_{\underline{n}_R=\underline{1}_r}^{\underline{N}_R} p_M(\underline{n}_R,\underline{N}_R)\ n_r$$

$$= \sum_{\underline{n}_R=\underline{1}_r}^{\underline{N}_R} n_r \frac{f_M(\underline{n}_R)}{G(\underline{N}_R)} g_{M-1}(\underline{N}_R-\underline{n}_R)$$

$$= \frac{1}{G(\underline{N}_R)} \sum_{\underline{n}_R=\underline{1}_r}^{\underline{N}_R} n_r \; f_M(\underline{n}_R) \; g_{M-1}(\underline{N}_R-\underline{n}_R)$$

But the multiple job class normalization constant is computed by

$$G(\underline{N}_R) = \sum_{\underline{n}_R=\underline{0}_r}^{\underline{N}_R} f_M(\underline{n}_R) \; g_{M-1}(\underline{N}_R-\underline{n}_R).$$

For an IS station it is easier to employ Little's result to compute the mean queue length of a station by job class. With an IS queueing discipline, the mean waiting time of a class r job at station i is simply s_{ir}. Thus, for an IS station,

$$\bar{n}_{ir}(\underline{N}_R) = X_{ir}(\underline{N}_R) \; s_{ir}.$$

2.3.6 MEAN WAITING TIME

To compute the mean waiting time of a class r job at a FCFS, PS, or LCFS station i, we can apply Litte's result. Hence,

$$\overline{w}_{ir}(\underline{N}_R) = \frac{\bar{n}_{ir}(\underline{N}_R)}{X_{ir}(\underline{N}_R)} .$$

The mean waiting time can also be written as a recursion if station i is load independent, viz.,

$$\overline{w}_{ir}(\underline{N}_R) = s_{ir}\ (1 + \bar{n}_i(\underline{N}_R - \underline{1}_r)).$$

This expression is obtained by substituting the load independent formulas for mean queue length and throughput in the previous equation. The recursive formula implies that the mean waiting time of a class r job at a load independent station equals its mean service time (s_{ir}) plus the mean waiting time spent by the previously queued jobs ($s_{ir}\ \bar{n}_i(\underline{N}_R - \underline{1}_r)$). This is one of the important results contained in [REIS78].

As mentioned in the previous section, the mean waiting time of a class r job at an IS station is s_{ir}.

2.3.7 MEAN RESPONSE TIME

The mean return time of a job to an IS station is also referred to as the station's mean response time. Let $T_{ir}(\underline{N}_R)$ denote the mean response time of a class r job at station i when there are $\underline{N}_R$ jobs in the closed network. By definition the mean response time of a class r job at station i is the average time it spends in the remainder of the network. Thus, applying Little's result to the remainder of the network,

$$T_{ir}(\underline{N}_R) = \frac{N_r - \bar{n}_{ir}(\underline{N}_R)}{X_{ir}(\underline{N}_R)}.$$

The term $N_r - \bar{n}_{ir}(\underline{N}_R)$ denotes the mean number of class r jobs in the remainder of the network.

2.4 MULTIPLE JOB CLASS CLOSED QUEUEING NETWORK MODELS (WITH JOB CLASS SWITCHING)

Once again we demonstrate that the above analysis techniques may be generalized, this time to handle closed queueing networks where jobs are permitted to change class membership. To treat this topic we must first modify our notation to accomodate this new feature.

Each job class has a separate set of transition frequencies governing the movement of jobs of that class between stations and into other classes. Specifically, a job of class r which completes service at station i will next require service at station j in class s with a certain frequency denoted $p_{(i,r)(j,s)}$ where $1 \leqq i,j \leqq M$ and $1 \leqq r,s \leqq R$. In section 2.3 we listed four conditions which a feasible state of the model must satisfy. The first three conditions still apply for this new model. But the fourth condition, which states that the sum of the number of class r jobs at all stations must be N_r, no longer holds. That is, the number of jobs in any class may no longer remain constant because jobs are permitted to change class membership while traversing the network.

However, a simple constraint does exist. The key idea is to study the transition frequency matrix $P=[p_{(i,r)(j,s)}]$ (see [MUNT73]). For convenience, define a stage of the network to be a pair (i,r) where i is a station index ($1 \leq i \leq M$) and r is a job class number ($1 \leq r \leq R$). The idea is to partition the set of job classes {1,2,...,R} into disjoint subsets such that a job which initially belongs to one class of a subset can only visit the other classes of the same subset. If jobs are not permitted to switch class, then each subset contains exactly one class. Since the partitioning guarantees that job classes in different subsets cannot communicate, then the total number of jobs in any subset must remain constant.

The procedure to identify these subsets requires two steps. First, we define the set of all classes which a class r job can be in as:

$$C_r = \{ s \mid \text{stage } (j,s) \text{ can be reached in zero or more transitions from stage } (i,r) \text{ for any } i,j \text{ such that } 1 \leq i,j \leq M\}, \text{ for } 1 \leq r \leq R \text{ and } 1 \leq s \leq R.$$

This actually defines R sets, one for each job class. The second step is to eliminate any duplicate sets and label the remaining U sets EC_1 EC_2, ... , EC_U[1]. Observe that $U \leq R$.

---

[1] The U here should not be confused with our prior use of U to denote utilization. It should be clear from context whether the U refers to utilization or to the number of equivalence classes. We apologize to the reader for this necessary but inconvenient double usage.

If, we let N'_r denote the initial number of jobs in class r (we used N_r previously), then

$$N_s = \sum_{r \in EC_s} N'_r \quad \text{for } s=1,2,\ldots,U$$

denotes the number of jobs belonging to "equivalence class s", (EC_s). Therefore, the fourth constraint in networks that do allow jobs to switch class membership is that N_s (for $1 \leqq s \leqq U$) must remain constant.

Now we can explicitly define the set of <u>all</u> feasible states to be

$$S(N_1, N_2, \ldots N_U, M) =$$

$$\{(\underline{n}_1, \underline{n}_2, \ldots, \underline{n}_i, \ldots, \underline{n}_M) \mid \underline{n}_i = (n_{i1}, \ldots, n_{ir}, \ldots, n_{iR}),\ n_{ir} \geqq 0,$$
$$\text{for } 1 \leqq i \leqq M \text{ and } 1 \leqq r \leqq R, \text{ and } \sum_{\substack{r \in EC_s \\ 1 \leqq i \leqq M}} n_{ir} = N_s \text{ for } 1 \leqq s \leqq U\}.$$

The total number of feasible states can be shown to be

$$\prod_{s=1}^{U} \begin{pmatrix} M \cdot |EC_s| + N_s - 1 \\ M \cdot |EC_s| - 1 \end{pmatrix}$$

where $|EC_s|$ is the cardinality of EC_s.

At this point an example will help illustrate and clarify these ideas. Consider a closed queueing network with three job classes (R=3) and two stations (M=2) each having a FCFS queueing discipline. Let the total number of jobs in the network be N=10; distribute these jobs among the classes as follows:

$$N'_1 = 2, \quad N'_2 = 3, \quad \text{and} \quad N'_3 = 5.$$

Let the transition frequency matrix be

$$P = \begin{array}{c} \\ (i,r) \\ (1,1) \\ (1,2) \\ (1,3) \\ (2,1) \\ (2,2) \\ (2,3) \end{array} \begin{array}{cccccc} & & (j,s) \rightarrow & & & \\ (1,1) & (1,2) & (1,3) & (2,1) & (2,2) & (2,3) \end{array}$$

(i,r)	(1,1)	(1,2)	(1,3)	(2,1)	(2,2)	(2,3)
(1,1)	1/2	0	0	0	0	1/2
(1,2)	0	0	0	0	1	0
(1,3)	0	0	3/4	0	0	1/4
(2,1)	1/4	0	0	3/4	0	0
(2,2)	0	1	0	0	0	0
(2,3)	1/2	0	0	0	0	1/2

Observe that jobs of class 1 switch to class 3 upon entering station 2 and that jobs of class 3 switch back to class 1 before entering station 1 ($p_{(1,1)(2,3)}=1/2$ and $p_{(2,3)(1,1)}=1/2$). Applying the procedure to detect equivalence classes, we have:

Step 1 (forming sets of communicating job classes):

$$C_1 = \{1,3\}, \quad C_2 = \{2\}, \text{ and } C_3=\{1,3\}$$

and

Step 2 (eliminating duplicate sets):

$$EC_1 = \{1,3\} \text{ and } EC_2 = \{2\}.$$

Thus, from the original three jobs classes (R=3), we have two equivalence classes (U=2). The number of jobs belonging to each job class is

$$N_1 = N'_1 + N'_3 = 2 + 5 = 7 \quad \text{and } N_2 = N'_2 = 2.$$

Finally, since $|EC_1| = 2$ and $|EC_2| = 1$, the total number of feasible states for this model is 480.

The equilibrium probability of the state $\underline{\underline{n}}=(\underline{n}_1,\underline{n}_2,\ldots,\underline{n}_i,\ldots,\underline{n}_M)$ is:

$$P(\underline{\underline{n}})=P(\underline{n}_1,\underline{n}_2,\ldots,\underline{n}_i,\ldots,\underline{n}_M)=\frac{1}{G(N_1,N_2,\ldots,N_U)}\prod_{i=1}^{M} f_i(\underline{n}_i)$$

where $f_i(\underline{n}_i)$ is the same function defined in section 2.3. The e_{ir} are the solutions of the following set of linear equations

$$e_{js} = \sum_{r=1}^{R}\sum_{i=1}^{M} e_{ir}\, p_{(i,r)(j,s)} \quad \text{for } 1\leqq s\leqq R \text{ and } 1\leqq j\leqq M.$$

$G(N_1,N_2,\ldots,N_U)$ is a normalizing constant chosen so that all the $P(\underline{\underline{n}})$ sum to one. All other quantities retain there former definitions.

2.4.1 COMPUTATION OF THE NORMALIZATION CONSTANT

In this section we describe an efficient iterative algorithm for the computation of the normalization constant for a multiple job class closed network in which jobs are permitted to switch class.

The first step is to note, as above, that the sum over all feasible states of the equilibrium state probabilities must be one. Hence,

$$1 = \sum_{\underline{\underline{n}}\in S(N_1,N_2,\ldots,N_U,M)} P(\underline{\underline{n}})$$

$$= \frac{1}{G(N_1, N_2, \ldots, N_U)} \sum_{\underline{n} \in S(N_1, N_2, \ldots, N_U, M)} \prod_{i=1}^{M} f_i(\underline{n}_i).$$

The following extension to our vector notation will be useful in many of the succeeding derivations:

$$\underline{N}_U \equiv (N_1, N_2, \ldots, N_U)$$

$$\underline{n}_U \equiv (n_1, n_2, \ldots, n_U)$$

$$\underline{k}_U \equiv (k_1, k_2, \ldots, k_U)$$

$$\underline{n}_U = \underline{k}_U \equiv n_1=k_1,\ n_2=k_2,\ \ldots,\ n_U=k_U$$

$$\underline{0}_U \equiv (0, 0, \ldots, 0)$$ (i.e., a vector of U zeroes)

$$\underline{1}_u \equiv (0, 0, \ldots, 1_u, \ldots, 0)$$ (i.e., a 1 in the uth component and U−1 zeroes)

$$\sum_{\underline{k}_U=\underline{0}_U}^{\underline{N}_U} \equiv \sum_{k_U=0}^{N_U} \cdots \sum_{k_2=0}^{N_2} \sum_{k_1=0}^{N_1}$$

$$\delta(\underline{n}_U) = \begin{cases} 0 & \text{if } n_1=n_2=\cdots=n_U=0 \\ 1 & \text{else} \end{cases}$$

By

$\sum_{\underline{n}_i=\underline{k}_U}$ where $\underline{n}_i$ is an R component vector, we mean

$$\sum_{\underline{n}_i \in \{(n_{i1}, n_{i2}, \ldots, n_{iR}) \mid \sum_{r \in EC_l} n_{ir} = k_l,\ l=1,2,\ldots,U\}}$$

Employing this vector notation, the normalization constant becomes

$$G(\underline{N}_U) = \sum_{\underline{\underline{n}} \in S(\underline{N}_U, M)} \prod_{i=1}^{M} f_i(\underline{n}_i) .$$

Notice that the structure of the normalization constant array is now given by the PASCAL-like declaration:

G : <u>array</u> [0..N_1,0..N_2,...,0..N_U] <u>of</u> <u>real</u>.

Thus, the array size is

$$\prod_{s=1}^{U} (N_s + 1) \quad \text{elements.}$$

To derive a computational algorithm for the normalization constant, we require the following auxiliary function:

$$g_m(\underline{n}_U) = \sum_{\underline{\underline{n}} \in S(\underline{n}_U, m)} \prod_{i=1}^{m} f_i(\underline{n}_i) .$$

Now for m>1,

$$g_m(\underline{n}_U) = \sum_{\underline{\underline{n}} \in S(\underline{n}_U, m)} \prod_{i=1}^{m} f_i(\underline{n}_i)$$

$$= \sum_{\underline{k}_U = \underline{0}_U}^{\underline{n}_U} \Big[\sum_{\substack{\underline{\underline{n}} \in S(\underline{n}_U, m) \\ \underline{n}_m = \underline{k}_U}} \prod_{i=1}^{m} f_i(\underline{n}_i) \Big]$$

$$= \sum_{\underline{k}_U=\underline{0}_U}^{\underline{n}_U} \left[\sum_{\underline{n}_m=\underline{k}_U} f_m(\underline{n}_m) \cdot \left[\sum_{\substack{\underline{\underline{n}}\in S(\underline{n}_U,m) \\ \underline{n}_m=\underline{k}_U}} \prod_{i=1}^{m-1} f_i(\underline{n}_i) \right]\right]$$

$$= \sum_{\underline{k}_U=\underline{0}_U}^{\underline{n}_U} \left[\sum_{\underline{n}_m=\underline{k}_U} f_m(\underline{n}_m) \cdot g_{m-1}(\underline{n}_U-\underline{k}_U) \right]$$

Since expressions similar to the last one will be encountered again, we prove the following lemma concerning such sums:

Lemma 3: The expression

$$\sum_{\underline{k}_U=\underline{0}_U}^{\underline{n}_U} \left[\sum_{\underline{n}_i=\underline{k}_U} f_i(\underline{n}_i) \cdot \varphi(\underline{k}_U) \right]$$

where φ is a function of the index variables $k_1, k_2, \ldots, k_U$ reduces to

$$\sum_{\underline{k}_U=\underline{0}_U}^{\underline{n}_U} fp_i(\underline{k}_U) \cdot \varphi(\underline{k}_U)$$

where

$$fp_i(\underline{k}_U) = \begin{cases} \dfrac{k_i!}{\prod_{k=1}^{k_i} \mu_i(k)} \prod_{s=1}^{U} \dfrac{1}{k_s!} \left[\sum_{r\in EC_s} (e_{ir}s_{ir}) \right]^{k_s} & \text{if station } i \text{ is FCFS load dependent} \\ k_i! \prod_{s=1}^{U} \dfrac{1}{k_s!} \left[\sum_{r\in EC_s} (e_{ir}s_{ir}) \right]^{k_s} & \text{if station } i \text{ is load independent FCFS,PS,LCFS} \\ \prod_{s=1}^{U} \dfrac{1}{k_s!} \left[\sum_{r\in EC_s} (e_{ir}s_{ir}) \right]^{k_s} & \text{if station } i \text{ has an IS service discipline} \end{cases}$$

and

$$k_i = \sum_{s=1}^{U} k_s$$

Proof: See Appendix A.

Applying Lemma 3, the auxiliary function becomes

$$g_m(\underline{n}_U) = \sum_{\underline{k}_U=\underline{0}_U}^{\underline{n}_U} fp_m(\underline{k}_U)\; g_{m-1}(\underline{n}_U-\underline{k}_U) \;.$$

When the service rate of the mth station is load dependent, this expression must be used. Once again, simplifications to this result exist when station m is either load independent or contains multiple identical servers.

To prove this, we require the following lemma.

Lemma 4: If station m is load independent and has a FCFS, PS, or LCFS service discipline, then

$$fp_m(\underline{n}_U) = \sum_{r=1}^{U} Y_{mr}\; fp_m(\underline{n}_U-\underline{1}_r)$$

where

$$Y_{mr} = \sum_{s\in EC_r} (e_{ms} s_{ms}) \;.$$

Proof: Similar to proof of Lemma 1.

Then when station m is load independent, and not an IS station,

$$g_m(\underline{n}_U) = \sum_{\underline{k}_U=\underline{0}_U}^{\underline{n}_U} fp_m(\underline{k}_U)\; g_{m-1}(\underline{n}_U-\underline{k}_U)$$

$$= g_{m-1}(\underline{n}_U) + \sum_{\underline{k}_U=\underline{0}_U}^{\underline{n}_U} \delta(\underline{k}_U)\ fp_m(\underline{k}_U)\ g_{m-1}(\underline{n}_U-\underline{k}_U)$$

$$= g_{m-1}(\underline{n}_U) + \sum_{r=1}^{U} Y_{mr}\ \left[\sum_{\underline{k}_U=\underline{0}_U}^{\underline{n}_U} \delta(\underline{k}_U) fp_m(\underline{k}_U-\underline{1}_r) g_{m-1}(\underline{n}_U-\underline{k}_U)\right]$$

$$= g_{m-1}(\underline{n}_U) + \sum_{r=1}^{U} Y_{mr}\ \left[\sum_{\underline{k}_U=\underline{1}_r}^{\underline{n}_U} fp_m(\underline{k}_U-\underline{1}_r)\ g_{m-1}(\underline{n}_U-\underline{k}_U) \right]$$

$$= g_{m-1}(\underline{n}_U) + \sum_{r=1}^{U} Y_{mr}\ \left[\sum_{\underline{k}_U=\underline{0}_U}^{\underline{n}_U-\underline{1}_r} fp_m(\underline{k}_U)\ g_{m-1}(\underline{n}_U-\underline{1}_r-\underline{k}_U) \right]$$

$$= g_{m-1}(\underline{n}_U) + \sum_{r=1}^{U} Y_{mr}\ g_m(\underline{n}_U-\underline{1}_r)$$

Hence,

$$g_m(n_1,n_2,\ldots,n_U) = g_{m-1}(n_1,n_2,\ldots n_U) + \sum_{r=1}^{U} Y_{mr}\ g_m(n_1,n_2,\ldots,n_r-1,\ldots,n_U) \qquad (2.13)$$

Equation 2.13 is a direct generalization of equation 2.8; there are only two differences. First, rather than having R job classes, equation 2.13 uses U "equivalence classes". Since $U \leqq R$, equation 2.13 may require less computation to evaluate than equation 2.8. We will return to this point later. Second, Y_{mr} has a slightly different meaning in equation 2.13, i.e.,

$$Y_{mr} = \sum_{s \in EC_r} e_{ms} s_{ms}$$

instead of

$$Y_{mr} = e_{mr} \; s_{mr}$$

as in equation 2.8.

Observe also that the proofs of equation 2.8 and 2.13 are almost identical. Aside from the aforementioned differences, we simply replaced $f_m(\underline{k}_R)$ with $fp_m(\underline{k}_U)$. The same observation will be used frequently henceforth.

In particular, it is straightforward to prove the following generalization of equation 2.9: if the mth station contains a_m multiple identical servers, then

$$g_m(\underline{n}_U) = g_{m-1}(\underline{n}_U) + \frac{1}{a_m} \sum_{r=1}^{U} Y_{mr} \Big[g_m(\underline{n}_U-\underline{1}_r) + \sum_{\substack{\underline{k}_U=\underline{0}_U \\ k_m \leqq a_m-2}}^{\underline{n}_U-\underline{1}_r} \left(\frac{a_m-k_m-1}{k_m+1} \right) fp_m(\underline{k}_U) \; g_{m-1}(\underline{n}_U-\underline{1}_r-\underline{k}_U) \Big] \tag{2.14}$$

where $k_m=k_1+k_2+\cdots+k_U$.

The initial condition for the recurrence relation of this section is

$$g_1(n_1,n_2,\ldots,n_U)=fp_1(n_1,n_2,\ldots,n_U) \quad \text{for} \quad \begin{cases} n_1=0,1,\ldots,N_1 \\ n_U=0,1,\ldots,N_U \end{cases}$$

2.4.2 MARGINAL QUEUE LENGTH DISTRIBUTION

The marginal probability of finding n_r jobs of class r $(r=1,2,\ldots,R)$ at station i is given by

$$p_i(\underline{n}_R,\underline{N}_U) = \sum_{\substack{\underline{\underline{n}}\in S(\underline{N}_U,M)\\ \underline{n}_i=\underline{n}_R}} P(\underline{n}_1,\underline{n}_2,\ldots,\underline{n}_{i-1},\underline{n}_R,\underline{n}_{i+1},\ldots,\underline{n}_M)$$

$$= \frac{f_i(\underline{n}_R)}{G(\underline{N}_U)} \sum_{\substack{\underline{\underline{n}}\in S(\underline{N}_U,M)\\ \underline{n}_i=\underline{n}_R}} \prod_{\substack{j=1\\ j\neq i}}^{M} f_j(\underline{n}_j)$$

Generalizing the auxiliary function definition to multiple job classes with class switching yields

$$g_m^{i}(\underline{n}_U) = \sum_{\substack{\underline{\underline{n}}\in S(\underline{N}_U,m)\\ \underline{n}_i=\underline{N}_U-\underline{n}_U}} \prod_{j=1}^{M} f_j(\underline{n}_j)$$

With this auxiliary function definition, the marginal distribution becomes

$$p_i(\underline{n}_R,\underline{N}_U) = \frac{f_i(\underline{n}_R)}{G(\underline{N}_U)} \; g_M^{i}(\underline{N}_U-\underline{k}_U)$$

where $\underline{k}_U=(k_1,k_2,\ldots,k_U)$ and

$$k_l = \sum_{r\in EC_l} n_{ir} \qquad \text{for } l=1,2,\ldots,U.$$

A straightforward generalization of the techniques of section 2.3 yield the following algorithms for computing the new auxiliary function:

if station i is load dependent:

$$g_M^{i}(\underline{n}_U) = G(\underline{n}_U) - \sum_{\underline{k}_U=\underline{0}_U}^{\underline{n}_U} \delta(\underline{k}_U)\; fp_i(\underline{k}_U)\; g_M^{i}(\underline{n}_U-\underline{k}_U) \qquad (2.15)$$

if station i is load independent:

$$g_M^{i}(\underline{n}_U) = G(\underline{n}_U) - \sum_{r=1}^{U} Y_{ir}\, G(\underline{n}_U-\underline{1}_r) \tag{2.16}$$

if station i contains a_i identical servers:

$$g_M^{i}(\underline{n}_U) = G(\underline{n}_U) - \frac{1}{a_i}\sum_{r=1}^{U} Y_{ir}\, [G(\underline{n}_U-\underline{1}_r) + \sum_{\substack{\underline{k}_U=\underline{0}_U \\ k_i \leqq a_i-2}}^{\underline{n}_U-\underline{1}_r} \left(\frac{a_i-k_i-1}{k_i+1}\right) fp_i(\underline{k}_U)\, g_M^{i}(\underline{n}_U-\underline{1}_r-\underline{k}_U) \tag{2.17}$$

2.4.3 THROUGHPUT

Let $X_{ir}(\underline{N}_U)$ denote the throughput of class r jobs at station i in a network with $\underline{N}_U$ jobs. In Appendix A we prove that

$$X_{ir}(\underline{N}_U) = e_{ir} \frac{G(\underline{N}_U-\underline{1}_q)}{G(\underline{N}_U)} \tag{2.18}$$

where q is the equivalence class of job class r.

This result applies for any of the four types of queueing disciplines. It also applies for stations with load dependent and load independent service rates!

2.4.4 UTILIZATION

Let $U_{ir}(\underline{N}_U)$ denote the utilization of station i by class r jobs in a network with $\underline{N}_U$ jobs. (This measure does not apply for an IS station).

For a load independent station, the multiple class utilization formula generalizes to the present case; in particular,

$$U_{ir}(\underline{N}_U) = s_{ir}\ X_{ir}(\underline{N}_U).$$

For a load dependent station, utilization must be calculated from its definition, i.e.,

$$U_{ir}(\underline{N}_U) = \sum_{\underline{n}_U=\underline{1}_q}^{\underline{N}_U} p_i(\underline{n}_R,\underline{N}_U)\ \frac{n_{ir}}{n_i}$$

2.4.5 MEAN QUEUE LENGTH

Let $\bar{n}_{ir}(\underline{N}_U)$ denote the average number of class r jobs at service center i in a network with $\underline{N}_U$ jobs. For a station with a FCFS, PS, or LCFS queueing discipline, the definition of mean queue length by job class is:

$$\bar{n}_{ir}(\underline{N}_U) = \sum_{\underline{n}_U=\underline{1}_q}^{\underline{N}_U} p_i(\underline{n}_R,\underline{N}_U)\ n_r$$

In Appendix A we prove that if station i is load independent, then

$$\bar{n}_{ir}(\underline{N}_U) = U_{ir}(\underline{N}_U)\ (1+\bar{n}_i(\underline{N}_U-\underline{1}_q))$$

where $\bar{n}_i(\underline{N}_U)$ denotes the <u>total</u> mean queue length at station i when there are $\underline{N}_U$ jobs in the network. This total mean queue length may be computed from

$$\bar{n}_{ir}(\underline{N}_U) = \sum_{r=1}^{R} U_{ir}(\underline{N}_U)\ (1+\bar{n}_i(\underline{N}_U-1_q))$$

with the initial condition $\bar{n}_i(\underline{0}_U)=0.0$.

The mean queue length of an IS station is computed by employing Little's result. With an IS queueing discipline, the mean waiting time of a class r job at station i is simply s_{ir}. Hence, for an IS station,

$$\bar{n}_{ir}(\underline{N}_U) = X_{ir}(\underline{N}_U)\ s_{ir}.$$

2.4.6 MEAN WAITING TIME

To compute the mean waiting time of a class r job at a FCFS, PS, or LCFS station i, we can apply Little's result. Thus,

$$\bar{w}_{ir}(\underline{N}_U) = \frac{\bar{n}_{ir}(\underline{N}_U)}{X_{ir}(\underline{N}_U)}$$

The mean waiting time can also be written as a recursion if station i is load independent, viz.,

$$\bar{w}_{ir}(\underline{N}_U) = s_{ir}\ (1+\bar{n}_i(\underline{N}_U-\underline{1}_q))$$

This expression is obtained by substituting the load

independent formulas for mean queue length and throughput in the previous equation.

As mentioned in the previous section, the mean waiting time of a class r job at an IS station is s_{ir}.

2.4.7 MEAN RESPONSE TIME

This performance measure applies only to IS stations. Let $T_{ir}(\underline{N}_U)$ denote the mean response time of a class r job at station i when there are $\underline{N}_U$ jobs in the closed network. By definition, the mean response time of a class r job at station i is the average time it spends in the remainder of the network. Thus, applying Little's result to the remainder of the network,

$$T_{ir}(\underline{N}_U) = \frac{N_r - \bar{n}_{ir}(\underline{N}_U)}{X_{ir}(\underline{N}_U)}$$

The term $N_r - \bar{n}_{ir}(\underline{N}_U)$ denotes the mean number of class r jobs in the remainder of the network.

2.5 CONCLUSIONS

In this chapter we have shown that the computational algorithms for closed queueing networks with only one job class may be extended to handle multiple job class networks. The multiple job class algorithms can be viewed as vector generalizations of the single class algorithms. For example, the single class throughput equation is

$$X_i(n) = e_i \frac{G(n-1)}{G(n)} ;$$

its multiple class counterpart is

$$X_{ir}(\underline{n}_R) = e_{ir} \frac{G(\underline{n}_R-\underline{1}_r)}{G(\underline{n}_R)} .$$

In addition, we have shown that the multiple class algorithms can be further extended to handle networks in which jobs are permitted to switch class. The performance measures of these networks are no more difficult to compute than for networks without class switching.

The main results of this chapter are summarized in Tables 2-3 through 2-17. Table 2-2 reviews the notation used in the chapter.

The next chapter describes the mean value analysis approach to solving queueing network models. Chapter 4 then discusses some practical problems arising when one tries to implement the algorithms in Chapters 2 and 3 in a computer program.

Table 2-2:
Notation of Chapter 2

Formal Symbolism	Meaning
a_i	Number of identical servers at station i
e_i	Relative arrival rate of jobs to station i
$f_i(k)$	Service function of station i with k jobs
$g_m(n)$	Auxiliary function for a network with m stations and n jobs
$g_m^i(n)$	Auxiliary function for a network with m-1 stations (excluding i) and n jobs
G(N)	Normalization constant for a network with M stations and N jobs
i	Station index ($1 \leqq i \leqq M$)
j	Station index ($1 \leqq j \leqq M$)
k	Job index ($0 \leqq k \leqq N$)
m	Station index
M	Number of stations in network
n	Job index ($0 \leqq n \leqq N$)
$\underline{n}$	State of network ($\underline{n}=(n_1, n_2, \ldots, n_i, \ldots, n_M)$)
n_i	Number of jobs at station i
$\bar{n}_i(N)$	Average queue length of ith station in a network containing N jobs
N	Number of jobs in closed network
p_{ij}	Transition frequency (fraction of departures from station i that proceed next to station j)

Table 2-2, Cont.:

Formal Symbolism	Meaning
$p_i(n,N)$	Marginal probability of finding exactly n jobs at station i (n=0,1,...,N)
$P(\underline{n})$	Equilibrium distribution of jobs in network
s_i	Mean service time at station i
$S(N,M)$	Set of feasible states in network with N jobs and M stations
$u_i(N)$	Mean number of active servers at station i
$U_i(N)$	Utilization of station i in N job network
$\overline{w}_i(N)$	Mean waiting time at station i
$X_i(N)$	Throughput of station i in N job network
Y_i	$Y_i=e_i s_i$
μ_i	Load independent service rate of station i
$\mu_i(k)$	Service rate of station with load of n jobs

Table 2-3:

Evaluation of normalization constant for a single class network with M stations and N jobs : G(N)

Auxiliary function	: $g_m(n)$ for $n=0,1,\ldots,N$ and $m=1,2,\ldots,M$
Invariant condition	: $g_m(0) = 1.0$ for all m, $1 \leqq m \leqq M$
Normalization constant	: $g_M(N) = G(N)$

Station m	$1 \leqq n \leqq N$	Algorithm	Where
load independent	$g_m(n) =$	m=1 : $Y_1 \; g_1(n-1)$	$Y_1 = e_1 s_1$
		else : $g_{m-1}(n) + Y_m \; g_m(n-1)$	$Y_m = e_m s_m$
load dependent	$g_m(n) =$	m=1 : $(e_1/\mu_1(n)) \; g_1(n-1)$	$f_m(0) = 1.0$
		else : $\sum_{k=0}^{n} f_m(k) \; g_{m-1}(n-k)$	$f_m(k) = \dfrac{e_m}{\mu_m(k)} f_m(k-1)$

Table 2-3, Cont.:

Station m	$1 \leq n \leq N$	Algorithm	Where
load dependent with a_m identical servers	$g_m(n) =$	$m=1 : \begin{cases} \frac{Y_1}{n} g_1(n-1) & \text{if } n \leq a_1 \\ \frac{Y_1}{a_1} g_1(n-1) & \text{if } n > a_1 \end{cases}$ $\text{else} : g_{m-1}(n) + \frac{Y_m}{a_m} \left[g_m(n-1) + \sum_{k=0}^{a_m-2} \left(\frac{a_m-k-1}{k+1} \right) f_m(k) g_{m-1}(n-1-k) \right]$	$Y_1 = e_1 s_1 = e_1/\mu_1(1)$ $a_1 =$ no. of identical servers at station 1

Table 2-4:

Evaluation of auxiliary function $g^i_M(n)$ for a single class network with M stations and N jobs

Auxiliary function	: $g^i_M(n)$ for $n=0,1,\ldots,N$ and $i=1,2,\ldots,M-1$
Invariant condition	: $g^i_M(0) = 1.0$ for $1 \leqq i \leqq M-1$
Special property	: $g^M_M(n) = g_{M-1}(n)$ for $n=0,1,\ldots,N$

Station i	$1 \leqq n \leqq N$ Algorithm	Where
load independent	$g^i_M(n) = G(n) - Y_i\ G(n-1)$	$Y_i = e_i s_i$
load dependent	$g^i_M(n) = G(n) - \sum_{k=1}^{n} f_i(k)\ g^i_M(n-k)$	$\begin{cases} f_i(0) = 1.0 \\ f_i(k) = \dfrac{e_i}{\mu_i(k)} f_i(k-1) \end{cases}$
load dependent with a_i identical servers	$g^i_M(n) = G(n) - \dfrac{Y_i}{a_i}\left[G(n-1) + \sum_{k=0}^{a_i-2}\left(\dfrac{a_i-k-1}{k+1}\right) f_i(k) g^i_M(n-1-k)\right]$	

<u>Table 2-5</u>:

Calculation of marginal queue length distribution for a single class network with M stations and N jobs : $p_i(n,N)$

Marginal Queue Length Distribution : $p_i(n,N)$ for $i=1,2,\ldots,M$ and $n=0,1,\ldots,N$		
<u>Station i</u>	<u>Algorithm</u>	<u>Where</u>
load independent	$p_i(n,N) = \dfrac{f_i(n)}{G(n)} [G(N-n) - Y_i\ G(N-1-n)]$	$\begin{cases} f_i(0) = 1.0 \\ f_i(n) = Y_i\ f_i(n-1) \\ Y_i = e_i s_i \end{cases}$
load dependent	$p_i(n,N) = \begin{cases} i \neq M : \dfrac{f_i(n)}{G(n)}\ g_M^i(N-n) \\ i = M : \dfrac{f_M(n)}{G(n)}\ g_{M-1}(N-n) \end{cases}$	$\begin{cases} f_i(0) = 1.0 \\ f_i(n) = \dfrac{e_i}{\mu_i(n)}\ f_i(n-1) \\ g_M^i(n) : \text{see Table 2-4} \end{cases}$

Table 2-6:

Calculation of throughput and utilization for a single class network with M stations and N jobs : $X_i(N)$ and $U_i(N)$

Throughput : $X_i(n)$ for $i=1,2,\dots,M$ and $n=1,2,\dots,N$
Utilization : $U_i(n)$ for $i=1,2,\dots,M$ and $n=1,2,\dots,N$

Station i	1≦n≦N Algorithm	Where
load independent or load dependent	$X_i(n) = e_i \dfrac{G(n-1)}{G(n)}$	e_i is from solution of job flow balance eqns.
load independent	$U_i(n) = \begin{cases} s_i X_i(n) \\ Y_i \dfrac{G(n-1)}{G(n)} \end{cases}$	s_i = mean service time at station i $Y_i = e_i s_i$
load dependent	$U_i(n) = \begin{cases} i \neq M : 1 - \dfrac{g_M^i(n)}{G(n)} \\ i = m : 1 - \dfrac{g_{M-1}(n)}{G(n)} \end{cases}$	$g_M^i(n)$: see Table 2-4 $g_{M-1}(n)$: see Table 2-3

Table 2-7:

Calculation of mean queue length for a single class network with M stations and N jobs : $\bar{n}_i(N)$

Mean queue length : $\bar{n}_i(n)$ for $i=1,2,\ldots,M$ and $n=1,2,\ldots,N$		
Station i	**$1 \leq n \leq N$** / **Algorithm**	**Where**
load independent	$\bar{n}_i(n) = \sum_{k=1}^{n} Y_i^k \frac{G(n-k)}{G(n)}$	$Y_i = e_i s_i$
	$\bar{n}_i(n) = U_i(n)(1+\bar{n}_i(n-1))$	$U_i(n)$: see Table 2-6
load dependent	$i \neq M$: $\bar{n}_i(n) = \sum_{k=1}^{n} k\, p_i(k,n)$	$p_i(k,n)$: see Table 2-5
	$i = M$: $\bar{n}_i(n) = \frac{1}{G(n)} \sum_{k=1}^{n} k f_M(k) g_{M-1}(n-k)$	evaluation takes place concurrently with that of $G(n)$

Table 2-8:

Evaluation of normalization constant for a multiple class network with M stations and $\underline{N}_R$ jobs : $G(\underline{N}_R)$

Auxiliary function : $g_m(\underline{n}_R)$ for $n_1=0,1,\ldots,N_1,\ldots,n_R=0,1,\ldots,N_R$ and $m=1,\ldots,M$

Normalization constant : $g_M(\underline{N}_R) = G(\underline{N}_R)$

Station m	$\underline{0}_R<\underline{n}_R\leqq\underline{N}_R$	Algorithm	Where
load independent	$g_m(\underline{n}_R)=$	m=1 : $\sum_{r=1}^{R} Y_{1r}\, g_1(\underline{n}_R-\underline{1}_r)$	$Y_{1r} = e_{1r}s_{1r}$
		else : $g_{m-1}(\underline{n}_R)+\sum_{r=1}^{R} Y_{mr}g_m(\underline{n}_R-\underline{1}_r)$	$Y_{mr} = e_{mr}s_{mr}$
load dependent	$g_m(\underline{n}_R)=$	m=1 : $\frac{1}{\mu_1(n_1)}\sum_{r=1}^{R} e_{1r}\, g_1(\underline{n}_R-\underline{1}_r)$	$n_1=n_{11}+n_{12}+\cdots+n_{1R}$
		else : $\sum_{\underline{k}_R=\underline{0}_R}^{\underline{n}_R} f_m(\underline{k}_R)\, g_{m-1}(\underline{n}_R-\underline{k}_R)$	$f_m(\underline{0}_R) = 1.0$; $f_m(\underline{k}_R)=\sum_{r=1}^{R} e_{mr}f_m(\underline{n}_R-\underline{1}_r)\frac{1}{\mu_m(k_m)}$

Table 2-8, Cont.:

Station m	$\underline{0}_R < \underline{n}_R \leq \underline{N}_R$ Algorithm / Where
load dependent with a_m identical servers	$$g_m(\underline{n}_R) = \begin{cases} m=1 : & \begin{cases} \dfrac{1}{n_1 \mu_1(1)} \sum_{r=1}^{R} e_{1r}\, g_1(\underline{n}_R - \underline{1}_r) & \text{if } n_1 \leq a_1 \\ & n_1 = n_{11} + n_{12} + \cdots + n_{1R} \\ \dfrac{1}{a_1 \mu_1(1)} \sum_{r=1}^{R} e_{1r}\, g_1(\underline{n}_R - \underline{1}_r) & \text{if } n_1 > a_1 \end{cases} \\ \text{else} : & \begin{cases} g_{m-1}(\underline{n}_R) + \dfrac{1}{a_m} \sum_{r=1}^{R} Y_{mr} \Bigg[g_m(\underline{n}_R - \underline{1}_r) \\ \sum_{\substack{\underline{k}_R = \underline{0}_R \\ k_m \leq a_m - 2}}^{\underline{n}_R - \underline{1}_r} \left(\dfrac{a_m - k_m - 1}{k_m + 1} \right) f_m(\underline{k}_R)\, g_{m-1}(\underline{n}_R - \underline{1}_r - \underline{k}_R) \Bigg] \end{cases} \end{cases}$$

Table 2-9:

Evaluation of auxiliary function $g^{i}_{M}(\underline{n}_R)$ for a multiple class network with M stations and $\underline{N}_R$ jobs

Auxiliary function	: $g^{i}_{M}(\underline{n}_R)$ for $n_1=0,1,..,N_1,...,n_R=0,1,..,N_R$ and $m=1,..,M$	
Invariant condition	: $g^{i}_{M}(\underline{0}_R) = 1.0$ for $1 \leqq i \leqq M-1$	
Station i	$\underline{0}_R < \underline{n}_R \leqq \underline{N}_R$ Algorithm	Where
load independent	$g^{i}_{M}(\underline{n}_R) = G(\underline{n}_R) - \sum_{r=1}^{R} Y_{ir}\, G(\underline{n}_R - \underline{1}_r)$	$Y_{ir} = e_{ir} s_{ir}$
load dependent	$g^{i}_{M}(\underline{n}_R) = G(\underline{n}_R) - \sum_{\underline{k}_R=\underline{0}_R}^{\underline{n}_R} \delta(\underline{k}_R) f_i(\underline{k}_R) g^{i}_{M}(\underline{n}_R - \underline{k}_R)$	
load dependent with a_i identical servers	$g^{i}_{M}(\underline{n}_R) = G(\underline{n}_R) - \frac{1}{a_i} \sum_{r=1}^{R} Y_{ir} \Big[G(\underline{n}_R - \underline{1}_r) + \sum_{\substack{\underline{k}_R=\underline{0}_R \\ k_i \leqq a_i-2}}^{\underline{n}_R-\underline{1}_r} \left(\frac{a_i - k_i - 1}{k_i + 1} \right) f_i(\underline{k}_R)\; g^{i}_{M}(\underline{n}_R - \underline{1}_r - \underline{k}_R) \Big]$	

Table 2-10:

Evaluation of marginal queue length distribution for a multiple class network with M stations and $\underline{N}_R$ jobs : $p_i(\underline{n}_R, \underline{N}_R)$

Marginal Queue Length Distribution : $p_i(\underline{n}_R, \underline{N}_R)$ for $i=1,2,\ldots,M$ and $\underline{0}_R \leqq \underline{n}_R \leqq \underline{N}_R$		
Station i	$\underline{0}_R \leqq \underline{n}_R \leqq \underline{N}_R$ Algorithm	Where
load independent	$p_i(\underline{n}_R, \underline{N}_R) = \dfrac{f_i(\underline{n}_R)}{G(\underline{N}_R)} \left[G(\underline{N}_R - \underline{n}_R) - \sum_{r=1}^{R} Y_{ir} \, G(\underline{N}_R - \underline{1}_r - \underline{n}_R) \right]$	
load dependent	$p_i(\underline{n}_R, \underline{N}_R) = \begin{cases} i \neq M : \dfrac{f_i(\underline{n}_R)}{G(\underline{N}_R)} \, g_M^i(\underline{N}_R - \underline{n}_R) \\ i = M : \dfrac{f_i(\underline{n}_R)}{G(\underline{N}_R)} \, g_{M-1}(\underline{N}_R - \underline{n}_R) \end{cases}$	$g_M^i(\underline{n}_R)$: see Table 2-9 $g_{M-1}(\underline{n}_R)$: see Table 2-8

Table 2-11:

Evaluation of throughput and utilization for a multiple class network with M stations and $\underline{N}_R$ jobs : $X_{ir}(\underline{N}_R)$ and $U_{ir}(\underline{N}_R)$

Throughput : $X_{ir}(\underline{n}_R)$ for $i=1,2,\ldots,M$ and $n_1=0,1,\ldots,N_1,\ldots,n_R=0,1,\ldots,N_R$

Utilization : $U_{ir}(\underline{n}_R)$ for $i=1,2,\ldots,M$ and $n_1=0,1,\ldots,N_1,\ldots,n_R=0,1,\ldots,N_R$

Station i	$\underline{0}_R \leqq \underline{n}_R \leqq \underline{N}_R$ Algorithm	Where
load independent or load dependent	$X_{ir}(\underline{n}_R) = e_{ir} \dfrac{G(\underline{n}_R-\underline{1}_r)}{G(\underline{n}_R)}$	e_{ir} is from solution of job flow balance eqns.
load independent	$U_{ir}(\underline{n}_R) = \begin{cases} s_{ir} X_{ir}(\underline{n}_R) \\ Y_{ir} \dfrac{G(\underline{n}_R-\underline{1}_r)}{G(\underline{n}_R)} \end{cases}$	s_{ir} = mean service time of station i by class r job $Y_{ir} = e_{ir}s_{ir}$
load dependent	$U_{ir}(\underline{n}_R) = \sum_{\underline{n}_R=\underline{1}_r}^{\underline{N}_R} p_i(\underline{n}_R,\underline{N}_R) \dfrac{n_{ir}}{n_i}$	$n_i=n_{i1}+\cdots+n_{iR}$

Table 2-12:

Evaluation of mean queue length for a multiple class network with M stations and $\underline{N}_R$ jobs : $\bar{n}_{ir}(\underline{N}_R)$

Mean queue length : $\bar{n}_{ir}(\underline{n}_R)$ for $i=1,2,\ldots,M$ and $n_1=0,1,\ldots,N_1,\ldots,n_R=0,1,\ldots,N_R$			
Station i	$\underline{0}_R<\underline{n}_R\leqq\underline{N}_R$	Algorithm	Where
load independent or load dependent	$\bar{n}_{ir}(\underline{n}_R)$	$= \sum_{\underline{k}_R=\underline{1}_r}^{\underline{n}_R} p_i(\underline{k}_R,\underline{n}_R)\ k_r$	$p_i(\underline{k}_R,\underline{n}_R)$: see Table 2-10
load independent	$\bar{n}_{ir}(\underline{n}_R)$	$= \sum_{r=1}^{R} U_{ir}(\underline{n}_R)\ (1+\bar{n}_i(\underline{n}_R-\underline{1}_r))$	$U_{ir}(\underline{n}_R)$: see Table 2-11
load independent	$\bar{n}_i(\underline{n}_R)$	$= \sum_{r=1}^{R} U_{ir}(\underline{n}_R)\ (1+\bar{n}_i(\underline{n}_R-\underline{1}_r))$	$\bar{n}_i(\underline{0}_R)=0.0$

Table 2-13:

Evaluation of normalization constant for a multiple class network with M stations and $\underline{N}_U$ jobs : $G(\underline{N}_U)$

Auxiliary function : $g_m(\underline{n}_U)$ for $n_1=0,1,\ldots,N_1,\ldots,n_U=0,1,\ldots,N_U$ and $m=1,\ldots,M$			
Normalization constant : $g_M(\underline{N}_U) = G(\underline{N}_U)$			
Station m	$\underline{0}_U<\underline{n}_U\leqq\underline{N}_U$	Algorithm	Where
load independent	$g_m(\underline{n}_U)=$	$m=1$: $\sum_{r=1}^{U} Y_{1r}\, g_1(\underline{n}_U-\underline{1}_r)$	$Y_{1r} = \sum_{s\in EC_r} e_{1s}s_{1s}$
		else : $g_{m-1}(\underline{n}_U)+\sum_{r=1}^{U} Y_{mr}g_m(\underline{n}_U-\underline{1}_r)$	$Y_{mr} = \sum_{s\in EC_r} e_{ms}s_{ms}$
load dependent	$g_m(\underline{n}_U)=$	$m=1$: $\frac{1}{\mu_1(n_1)}\sum_{r=1}^{U} e_{1r}\, g_1(\underline{n}_U-\underline{1}_r)$	$n_1=n_{11}+n_{12}+\cdots+n_{1U}$
		else : $\sum_{\underline{k}_U=\underline{0}_U}^{\underline{n}_U} fp_m(\underline{k}_U)\; g_{m-1}(\underline{n}_U-\underline{k}_U)$	$fp_m(\underline{0}_U) = 1.0$; $fp_m(\underline{k}_U)= \sum_{r=1}^{U} e_{mr}fp_m(\underline{n}_U-\underline{1}_r)\,\frac{1}{\mu_m(k_m)}$

Table 2-13, Cont.:

Station m	$\underline{0}_U < \underline{n}_U \leq \underline{N}_U$	Algorithm	Where
load dependent with a_m identical servers	$g_m(\underline{n}_U) =$	$m=1$: $\dfrac{1}{n_1 \mu_1(1)} \sum_{r=1}^{U} e_{1r}\, g_1(\underline{n}_U - \underline{1}_r)$	if $n_1 \leq a_1$
			$n_1 = n_{11} + n_{12} + \cdots + n_{1U}$
		$m=1$: $\dfrac{1}{a_1 \mu_1(1)} \sum_{r=1}^{U} e_{1r}\, g_1(\underline{n}_U - \underline{1}_r)$	if $n_1 > a_1$
		else : $g_{m-1}(\underline{n}_U) + \dfrac{1}{a_m} \sum_{r=1}^{U} Y_{mr} \Big[g_m(\underline{n}_U - \underline{1}_r) + \sum_{\substack{\underline{k}_U = \underline{0}_U \\ k_m \leq a_m - 2}}^{\underline{n}_U - \underline{1}_r} \left(\dfrac{a_m - k_m - 1}{k_m + 1} \right) fp_m(\underline{k}_U)\, g_{m-1}(\underline{n}_U - \underline{1}_r - \underline{k}_U) \Big]$	

Table 2-14:

Evaluation of auxiliary function $g^i_M(\underline{n}_U)$ for a multiple class network with M stations and $\underline{N}_U$ jobs

Auxiliary function	: $g^i_M(\underline{n}_U)$ for $n_1=0,1,..,N_1,..,n_U=0,1,..,N_U$ and $m=1,..,M$	
Invariant condition	: $g^i_M(\underline{0}_U) = 1.0$ for $1 \leqq i \leqq M-1$	
Station i	$\underline{0}_U < \underline{n}_U \leqq \underline{N}_U$ Algorithm	Where
load independent	$g^i_M(\underline{n}_U) = G(\underline{n}_U) - \sum_{r=1}^{U} Y_{ir}\, G(\underline{n}_U - \underline{1}_r)$	$Y_{ir} = \sum_{s \in EC_r} e_{is} s_{is}$
load dependent	$g^i_M(\underline{n}_U) = G(\underline{n}_U) - \sum_{\underline{k}_U=\underline{0}_U}^{\underline{n}_U} \delta(\underline{k}_U) fp_i(\underline{k}_U) g^i_M(\underline{n}_U - \underline{k}_U)$	
load dependent with a_i identical servers	$g^i_M(\underline{n}_U) = G(\underline{n}_U) - \frac{1}{a_i} \sum_{r=1}^{U} Y_{ir} \Big[G(\underline{n}_U - \underline{1}_r) + \sum_{\substack{\underline{k}_U=\underline{0}_U \\ k_i \leqq a_i-2}}^{\underline{n}_U-\underline{1}_r} \left(\frac{a_i - k_i - 1}{k_i + 1} \right) fp_i(\underline{k}_U)\; g^i_M(\underline{n}_U - \underline{1}_r - \underline{k}_U) \Big]$	

Table 2-15:

Evaluation of marginal queue length distribution for a multiple class network with M stations and $\underline{N}_U$ jobs : $p_i(\underline{n}_R,\underline{N}_U)$

Marginal Queue Length Distribution : $p_i(\underline{n}_R,\underline{N}_U)$ for $i=1,2,\ldots,M$ and $\underline{0}_R \leq \underline{n}_R \leq \underline{N}_U$		
Station i	$\underline{0}_R \leq \underline{n}_R \leq \underline{N}_U$ Algorithm	Where
load independent	$p_i(\underline{n}_R,\underline{N}_U) = \dfrac{f_i(\underline{n}_R)}{G(\underline{N}_U)} [G(\underline{N}_U-\underline{k}_U) - \sum_{r=1}^{U} Y_{ir} G(\underline{N}_U-\underline{1}_r-\underline{k}_U)]$	$k_l = \sum_{r \in EC_l} n_{ir}, \quad l=1,\ldots,U$
load dependent	$p_i(\underline{n}_R,\underline{N}_U) = \begin{cases} i \neq M : \dfrac{f_i(\underline{n}_R)}{G(\underline{N}_U)} g_M^i(\underline{N}_U-\underline{k}_U) \\ i=M : \dfrac{f_i(\underline{n}_R)}{G(\underline{N}_U)} g_{M-1}(\underline{N}_U-\underline{n}_U) \end{cases}$	$g_M^i(\underline{n}_U)$: see Table 2-14 $g_{M-1}(\underline{n}_U)$: see Table 2-13

Table 2-16:

Evaluation of throughput and utilization for a multiple class network with M stations and $\underline{N}_U$ jobs : $X_{ir}(\underline{N}_U)$ and $U_{ir}(\underline{N}_U)$

Throughput : $X_{ir}(\underline{n}_U)$ for $i=1,2,\ldots,M$ and $n_1=0,1,\ldots,N_1,\ldots,n_U=0,1,\ldots,N_U$

Utilization : $U_{ir}(\underline{n}_U)$ for $i=1,2,\ldots,M$ and $n_1=0,1,\ldots,N_1,\ldots,n_U=0,1,\ldots,N_U$

Station i	$\underline{0}_U \leqq \underline{n}_U \leqq \underline{N}_U$ Algorithm	Where
load independent or load dependent	$X_{ir}(\underline{n}_U) = e_{ir} \dfrac{G(\underline{n}_U-\underline{1}_q)}{G(\underline{n}_U)}$	q is the equivalence class of job class r
load independent	$U_{ir}(\underline{n}_U) = \begin{cases} s_{ir}\, X_{ir}(\underline{n}_U) \\ Y_{ir} \dfrac{G(\underline{n}_U-\underline{1}_q)}{G(\underline{n}_U)} \end{cases}$	s_{ir} = mean service time of station i by class r job $Y_{ir} = \sum_{s \in EC_r} (e_{ms} s_{ms})$
load dependent	$U_{ir}(\underline{n}_U) = \sum_{\underline{n}_U=\underline{1}_q}^{\underline{N}_U} p_i(\underline{n}_r, \underline{N}_U) \dfrac{n_{ir}}{n_i}$	$n_i = n_{i1} + \cdots + n_{iR}$

Table 2-17:

Evaluation of mean queue length for a multiple class network with M stations and $\underline{N}_U$ jobs : $\bar{n}_{ir}(\underline{N}_U)$

Mean queue length : $\bar{n}_{ir}(\underline{n}_U)$ for $i=1,2,\ldots,M$ and $n_1=0,1,\ldots,N_1,\ldots,n_U=0,1,\ldots,N_U$

Station i	$\underline{0}_U<\underline{n}_U\leqq\underline{N}_U$		Algorithm	Where
load independent or load dependent	$\bar{n}_{ir}(\underline{n}_U)$	=	$\sum_{\underline{k}_R=\underline{1}_q}^{\underline{n}_U} p_i(\underline{k}_R,\underline{n}_U)\ k_r$	$p_i(\underline{k}_R,\underline{n}_U)$: see Table 2-10
load independent	$\bar{n}_{ir}(\underline{n}_U)$	=	$\sum_{r=1}^{R} U_{ir}(\underline{n}_U)\ (1+\bar{n}_i(\underline{n}_U-\underline{1}_q))$	$U_{ir}(\underline{n}_U)$: see Table 2-11
load independent	$\bar{n}_i(\underline{n}_U)$	=	$\sum_{r=1}^{R} U_{ir}(\underline{n}_U)\ (1+\bar{n}_i(\underline{n}_U-\underline{1}_q))$	$\bar{n}_i(\underline{0}_R)=0.0$

CHAPTER 3

PRACTICAL CONSIDERATIONS

3.1 INTRODUCTION

In the last chapter we analyzed the computational aspects of closed queueing networks with multiple job classes. The assortment of formulae for computing the performance measures of these networks is bewildering. This chapter discusses some practical problems arising in the design and implementation of a program to solve this class of queueing network models.

Conceptually, a computer program that evaluates the performance measures of a queueing network model can be divided into four modules (see Figure 3-1). The user interface module receives a description of the model to be evaluated. The complexity of this module depends to a large

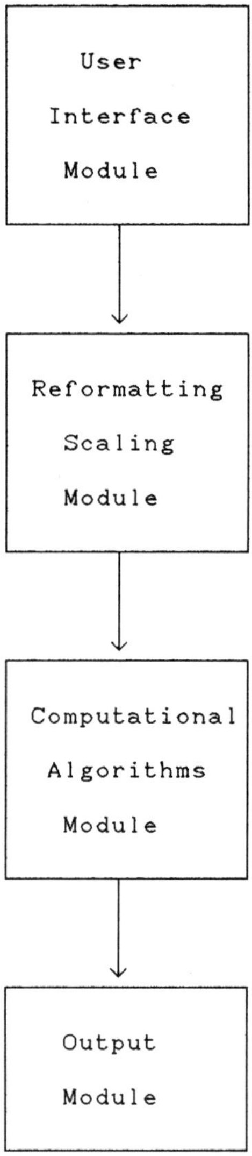

Figure 3-1:
Program Modules

extent on the sophistication of the intended users of the program. If the users are familiar with queueing network models, this module may accept simply a terse description of the network. For less knowledgeable users the approach adopted by BEST/1 is recommended[1] [BUZE78b]. The user interface module of BEST/1 prompts the user for a description of the network and at each step checks the validity of the input; it also accepts commands selecting the types of reports to be returned to the user once the model is evaluated. In BEST/1, about 90% of the code is in the user interface module; in the package described here (PNET -- the Purdue Network-of-Queues evaluator) only about 20% of the code is in this module.

Once the model's description has been input, control is transferred to the reformatting module. It determines the configuration of the network best suited for the efficient evaluation of performance measures by the computational algorithms. Various reformatting schemes are discussed in Section 3.2. This module is also responsible for scaling the input parameters to avoid arithmetic overflow and/or underflow in the computational algorithms (see Section 3.3).

---

[1] BEST/1, a commercial queueing network evaluator, is a proprietary product of BGS Systems, INC.

The third module computes the performance measures of the reformatted network using the algorithms developed in Chapter 2. These performance measures are then reordered to conform to the original network description and output via the fourth module.

There are several ways to implement a program to solve multiple class queueing network models. In the next chapter we discuss an implementation recently proposed by Reiser and Lavenberg [REIS78]. Their clever method avoids computation of the normalization constant, but some of their recursive algorithms may not be numerically stable [CHAN79]. In this chapter we discuss our own package (PNET) for solving multiple class queueing network models.

3.2 ELIMINATING MULTIPLE INFINITE SERVER STATIONS

In Chapter 2 (section 2.2.1) we observed that a queueing network with multiple IS stations could cause computational difficulties since the inefficient load dependent algorithms would be invoked for every IS station except the first. Fortunately, it is simple to avoid this extra computation. The idea is to coalesce all IS stations into one by incorporating new job classes and using the job class switching feature of BCMP networks.

Recall first that an IS station may service many different job classes, each having different mean service times. For simplicity we restrict our attention to a single class network with n distinct IS stations. Without loss of generality assume that these IS stations are indexed 1 to n with respective mean service times s_1, s_2, ... , s_n. These n IS stations can be represented by one IS station having n job class dependent mean service times. A simple example will help clarify this idea.

Consider a closed queueing network with three stations and one job class. Let the network contain two IS stations and one FCFS station arranged as depicted in Figure 3-2. The mean service times of the stations are s_1, s_2, and s_3. Notice that a job leaving station 2 enters station 3 with a transition frequency of q_{23}. Upon entering station 3 the job experiences a mean delay (service time) s_3 before returning to station 2. We can transform this network into an "equivalent" two station network by adding a new job class as shown in Figure 3-3. A job leaving station 2 enters IS station 1 in either its original job class 1 or a new job class denoted 2. Job class 1 has a mean service time s_{11} ($=s_1$) and job class 2 has a mean service time s_{12} ($=s_3$). Finally, a job departing station 1 in class 2 switches back to class 1 before entering station 2.

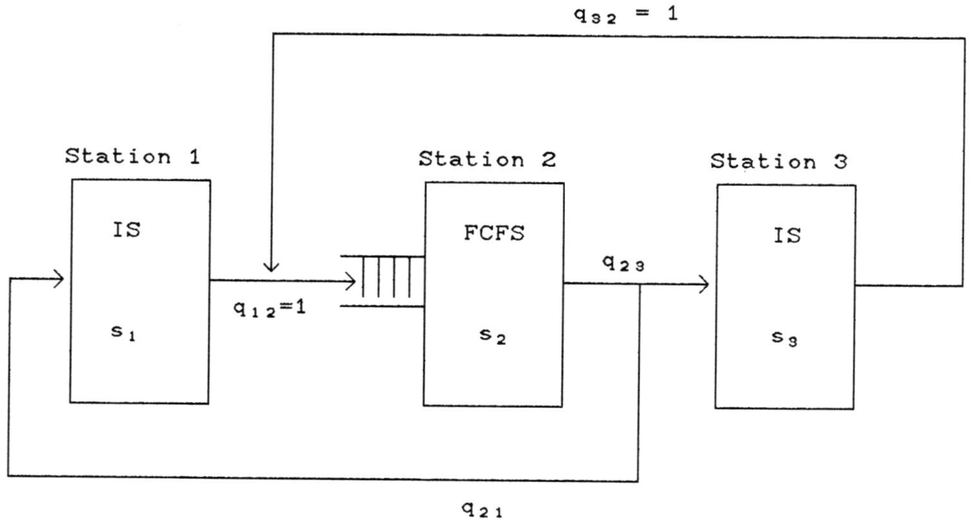

Figure 3-2:

Multiple IS station network

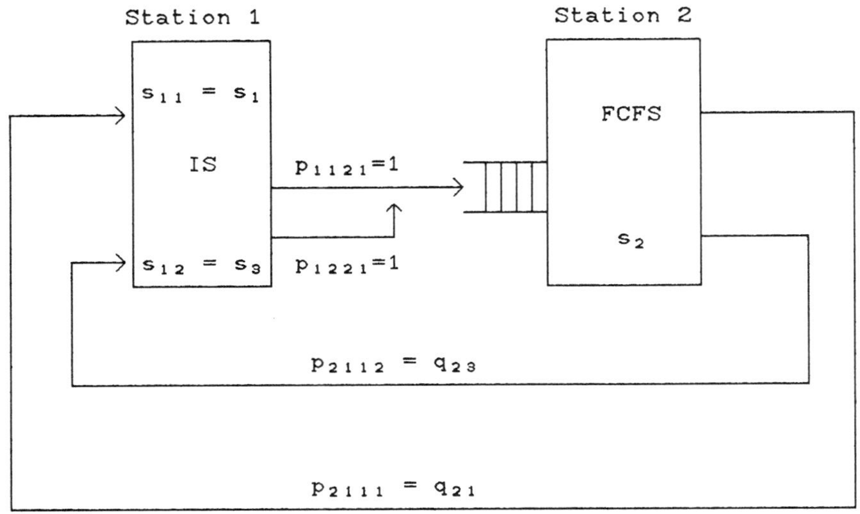

<u>Figure 3-3</u>:

Restructured network eliminating IS station

Table 3-1:

Parameters for original multiple IS station network

R = 1 class

M = 3 stations

N = 5 jobs

Station	Queueing Discipline	Mean Service Time (msec)
1	IS	10
2	FCFS	15
3	IS	20

$q_{12} = q_{32} = 1$

$q_{23} = 0.4$

$q_{21} = 0.6$

Table 3-2:
Parameters for restructured network eliminating IS station

$R = 2$ class

$U = 1$ equivalence class

$M = 2$ stations

$N_1 = 5$ jobs in class 1 and $N_2 = 0$ jobs in class 2

Station	Queueing Discipline	Mean Service Time Class 1	Mean Service Time Class 2
1	IS	10	20
2	FCFS	15	—

$p_{(1,1)(2,1)} = q_{12} = 1$

$p_{(1,2)(2,1)} = q_{32} = 1$

$p_{(2,1)(1,1)} = q_{21} = 0.6$

$p_{(2,1)(1,2)} = q_{23} = 0.4$

Table 3-3:
Performance measures for original multiple IS network

Station	Mean Queue Length	Throughput	Utilization
1	0.3991	0.0399	—
2	4.0688	0.0665	0.9977
3	0.5321	0.0266	—

Table 3-4:
Performance measures for restructured network eliminating IS station

	Mean Queue Length		Throughput		
Station	Class 1	Class 2	Class 1	Class 2	Util.
1	0.3991	0.5321	0.0399	0.0266	—
2	4.0688	—	0.0665	—	0.9977

The performance measures of the restructured network are exactly the same as the original network. Tables 3-1 through 3-4 present a numerical example which illustrates the equivalence.

The proof of this result is outlined below. We must show that the arrivals and departures from the coalesced IS station simulate those in the original network.

1. Arrival. In the original network, a job comes to IS_i; in the modified network to the IS station indexed 1 where it enters class i.

2. Departure. In the original network, a job moves from IS_i to station j with probability p_{ij}; in the modified network it moves from the IS station indexed 1 in class i to station j with probability $p_{(1,i)(j,1)}=p_{ij}$.

This proof can be generalized to handle multiple job class models with multiple IS stations.

3.3 SCALING

Reiser and Kobayshi [REIS75] have demonstrated that a careful scaling of the relative throughputs, e_{ir}, can avoid overflow and/or underflow during the computation of the normalization constant. This is because the quantities $Y_{ir}=e_{ir}s_{ir}$ raised to powers are required to compute the f_i, which in turn are needed to evaluate the normalization constant (see Chapter 2).

We illustrate the problem by the following example. Consider a network with parameters displayed in Table 3-5.

Table 3-5:

Parameters of example network

$R = 2$ classes,

$M = 3$ stations

Station	Queueing Discipline	Mean Service Time (msec) Class 1 (s_{i1})	Class 2 (s_{i2})
1	PS	10	20
2	FCFS	15	15
3	FCFS	12	12

$p_{(1,1)(2,1)} = p_{(1,2)(2,2)} = 0.3$

$p_{(1,1)(3,1)} = p_{(1,2)(3,2)} = 0.7$

$p_{(1,1)(1,1)} = p_{(2,2)(1,2)} = 1.0$

$p_{(3,1)(1,1)} = p_{(3,2)(1,2)} = 1.0$

One solution of the job flow balance equations is

$$e_{11} = e_{12} = 1.0, \quad e_{21} = e_{22} = 0.3, \quad e_{31} = e_{32} = 0.7.$$

With this choice of the e_{ir}, the $Y_{ir}=e_{ir}s_{ir}$ become

$$Y_{11} = 10, \; Y_{12} = 20, \; Y_{21} = Y_{22} = 4.5, \; Y_{31} = Y_{32} = 8.4.$$

If there are $N_1=8$ jobs of class 1 and $N_2=5$ jobs of class 2, then the G matrix can be computed using the algorithms of the previous chapter. Table 3-6 presents the exponents of the numbers in the last column of the last G matrix. It is easy to see that the numbers are growing quite large even for a problem with a small number of jobs per class.

In order to control the magnitude of these numbers, we use the following scaling of the e_{ir} and hence the Y_{ir}. A scaling constant, C, is chosen such that the sum of the squares of the difference between the Y_{ir} and one is minimized. A straightforward derivation yields the scaling constant

$$C = \left(\sum_{r=1}^{R} \sum_{i=1}^{M} Y_{ir} \right) / \left(\sum_{r=1}^{R} \sum_{i=1}^{M} Y_{ir}^{2} \right).$$

This scaling constant is not necessarily optimal for controlling underflow/overflow in the computation of the normalization constant, but it is quite satisfactory for the cases we have studied.

For the above example, the scaling constant is C=0.082. The Y_{ir} are then

$$Y_{11}=0.82 \ , \ Y_{12}=1.64 \ , \ Y_{21}=Y_{22}=0.37 \ , \ Y_{31}=Y_{32}=0.69.$$

Table 3-7 displays the exponents of the quantities in the last column of the last G matrix, when the above scaled Y_{ir} are used in the computational algorithms. It is obvious from a comparison of Tables 3-6 and 3-7 that the simple scaling rule has substantially reduced the rate of growth of

Table 3-6:

Exponents of unscaled values in normalization constant array

	G_3	$G_3(N_1,N_2)$ for $N_2=5$
	0	6
	1	8
	2	10
N_1 ↓	3	11
	4	13
	5	14
	6	15
	7	17
	8	18

Table 3-7:

Exponents of scaled values in normalization constant array

	G_3	$G_3(N_1,N_2)$ for $N_2=5$
	0	1
	1	2
	2	2
N_1 ↓	3	3
	4	3
	5	3
	6	3
	7	4
	8	4

the numbers used in the computation of the normalization constant, thereby reducing the risk of overflow.

3.4 PNET PROGRAM IMPLEMENTATION

In this section we discuss the basic organization of our program which solves multiple job class queueing network models.

First, we must decide what information is required to describe a multiple class model. But, more importantly, we must input this description to the program in a convenient form. To describe a model we must at least input:

(·) The total number of stations in the network (M)

(·) The number of distinct job classes (R)

(·) The number of jobs in each job class N_1, N_2, ... , N_R

(·) For each of the M stations:

The queueing discipline of the station: one of FCFS, PS, IS, or LCFS;

The number of servers at the station;

The set of transition probabilities associated with the station ($p_{(ir)(js)}$);

The mean service time of each class of job (s_{ir}).

Using the transition probabilities input above, the program determines the number of equivalence classes (U). Then a particular solution of the flow balance equations:

$$e_{js} = \sum_{\substack{r \in EC_k \\ 1 \leq i \leq M}} e_{ir}\, p_{(i,r)(j,s)} \qquad s \in EC_k \text{ for } k=1,2,\ldots,U \text{ and } 1 \leq j \leq M$$

is determined. These numbers are then scaled using the scaling rule described in section 3.3.

Next, the normalization constant is computed. This is done by applying the iterative algorithms listed in Table 2-8 or 2-13. These tables also describe how the G matrix is initialized.

Once the normalization constant has been computed, it is a straightforward task to compute the performance measures of the Mth station (cf. Tables 2-10 through 2-12 or Tables 2-15 through 2-17). To compute the performance measures of the other stations, we first compute the auxiliary function

$$g_M^i(\underline{n}_R) \quad \text{or} \quad g_M^i(\underline{n}_U)$$

using the algorithms listed in Table 2-9 or 2-14 and then apply the tabulated algorithms to compute the performance measures.

The order in which stations are indexed is important with respect to the speed of the computational algorithms. If there are any IS stations, we merge them using the technique discussed in Section 3.2; then we index the resulting station 1. (This is because evaluation of the auxiliary function is not required to compute the performance measures of an IS station (see Chapter 3)). Otherwise label load dependent stations with 1 and M before using other indices. (If the load dependent station is indexed M, we avoid the evaluation of the auxiliary function using the load dependent algorithms. If the load dependent station is indexed 1, an intermediate step in the computation of the normalization constant can avoid the load dependent algorithms).

To implement the computation of the normalization constant, we require an array G of R dimensions. One possible declaration would be

G : <u>array</u> [0..N_1,0..N_2,...,0..N_R] <u>of</u> <u>real</u>;

There are two disadvantages to implementing the G array as an R dimensional structure. First, we do not know until <u>execution</u> time the number of job classes (R) in the network. We would like to avoid the restrictions imposed on the size of R by declaring the number of dimensions of the G array at compile time. Second, we do not know until execution time the number of jobs per class (N_1 , N_2 , ... , N_R). Again, we would like to avoid compile time restrictions on the values of the N_r. To overcome these problems, we simply declare the G array as a simple vector:

G : <u>array</u> [0..maxsize] <u>of</u> <u>real</u>;

Then we can implement an R dimensional structure by computing the appropriate location into this singly dimensioned array.

Another related problem is implementing summations such as:

$$\sum_{\underline{n}_R=\underline{0}_R}^{\underline{N}_R} \equiv \sum_{n_R=0}^{N_R} \cdots \sum_{n_2=0}^{N_2} \sum_{n_1=0}^{N_1}$$

The naive approach would use a set of nested for loops:

```
for n1:=0 to N1 do
    for n2:=0 to N2 do
        .
          .
            for nR:=0 to NR do
```

This solution is inadequate because the number of nested for loops imposes a compile time restriction on R (the number of job classes).

Our solution is to simulate the operation of the nested summation using a repeat-until construct as follows:

```
n:=zero;                    (*initialize vector n to zero*)
repeat
  .
  .
  .
until ¬sum(n,N,R);
```

where

```
vector      = array [1..maxr] of integer;
n,N,zero    : vector;
```

and the function sum is defined:

```
function sum(var lower : vector; upper : vector;
   max : integer) : boolean;
var
   flag    : boolean;
   i       : integer;
begin
   i:=1;
   repeat
      lower[i]:=lower[i]+1;
      flag:=true;
      if lower[i] > upper[i] then
      begin
         lower[i]:=0;
         i:=i+1;
         flag:=false;
      end;
   until ((i > max) or flag);
   sum:=i≤max;
end; (*of sum*)
```

3.5 SUMMARY

In this chapter we have discussed the basic organization of queueing network evaluator programs. They contain more than just a straightforward coding of the computational algorithms discussed in Chapter 2. These programs are usually tailored to the needs of the intended users. The BEST/1 package developed by BGS Systems, Inc., was written with the novice user in mind. Approximately 90% of the code is devoted to the user interface and output modules. Our own implementation was written for the sophisticated user. The user interface and output modules are less flexible than those of BEST/1 and comprise only 20% of the code.

This chapter also described various techniques for reformatting queueing networks to obtain the maximum efficiency in the evaluation of the performance measures. Without these techniques certain seemingly trivial queueing networks could become computationally infeasible to evaluate.

CHAPTER 4

COMPUTATIONAL ASPECTS OF CLOSED QUEUEING NETWORKS WITH MULTIPLE JOB CLASSES: THE MEAN VALUE ANALYSIS APPROACH

4.1 INTRODUCTION

In the last two chapters we studied computational algorithms for solving BCMP closed queueing networks based on the normalization constant approach. In this chapter we will be studying the mean value analysis approach to solving queueing network models. This technique was developed by Reiser and Lavenberg in [REIS78]. We extend their results to include exactly the same class of networks which we studied in Chapters 2 and 3; i.e., multiple class networks in which jobs are permitted to switch class.

Mean value analysis [REIS78] computes the standard performance measures of BCMP models beginning with two principles:

(1) The queue length distribution observed by a job upon its arrival to a given station is the same as the overall distribution seen by an outside observer when one less job is in the system.

(2) Little's law can be used to transform queue lengths and waiting times throughout the network.

The first principle is well known for special cases (e.g., the machine repairman system [COOP72]). Recently Sevcik and Mitrani [SEVC79] and Lavenberg and Reiser [LAVE79] have proved that this same property holds precisely in all product form queueing networks.

Mean value analysis is easy to understand from an intuitive point of view -- a statement which can not be made for the corresponding normalization constant method. In the next section we motivate the mean value analysis approach by studying its application to a central server model with load independent stations and a single class of jobs.

4.2 SINGLE CLASS NETWORKS WITH LOAD INDEPENDENT PARAMETERS

Consider a central server network with M load independent stations. The mean service time per request of station i is s_i. The frequency with which a job requests service from the ith station after completing a CPU processing interval is p_i. A job making a CPU to CPU transition is regarded as having left the system and having been immediately replaced by another; thus the flow along this path represents the system throughput. Figure 4-1 depicts this model and its associated parameters. (We assume all stations have a FCFS queueing discipline.)

From these parameters the performance measures of this queueing network can be computed using the methods developed in Chapter 2. The solution steps are:

(1) Solve the job flow balance equations for the e_i. It is easy to show that $e_1=1$ and $e_i=p_i$ for $i=2,\ldots,M$ is one particular solution of these equations.

(2) Compute $Y_i=e_i s_i$ for $i=1,2,\ldots,M$.

(3) Let there be N jobs in the network. Evaluate the normalization constant by employing Buzen's algorithm (cf. eq. 2.2 and Figure 2-1).

(4) Compute the performance measures of the network:

$$X_i(n),\ U_i(n),\ \bar{n}_i(n),\ \text{and}\ \bar{w}_i(n)$$

for $i=1,2,\ldots,M$ and $n=1,2,\ldots,N$.

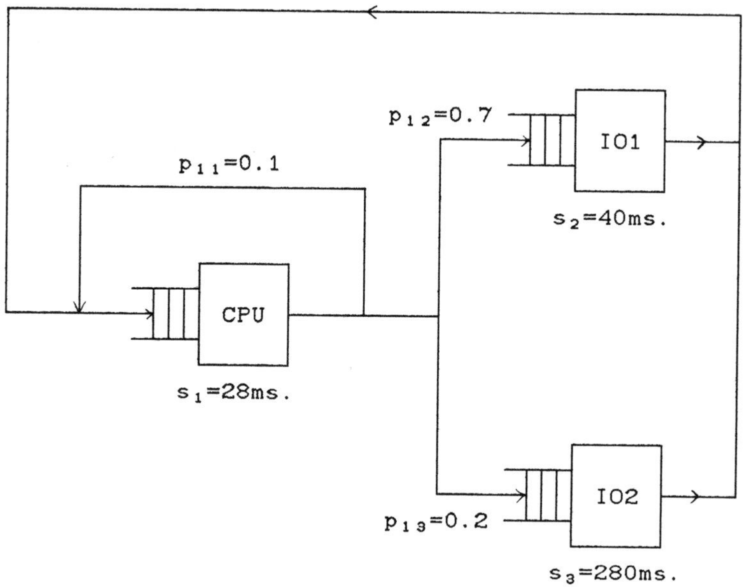

Figure 4-1: Central Server Network

We now demonstrate that the evaluation of the normalization constant (step 3) is unnecessary. First, carefully examine the load independent formulas listed below for evaluating the performance measures (cf. Chapter 2):

Throughput : $X_i(n) = e_i \dfrac{G(n-1)}{G(n)}$

Utilization : $U_i(n) = s_i \; X_i(n)$

Mean queue length : $\bar{n}_i(n) = X_i(n) \; \bar{w}_i(n)$
$\bar{n}_i(0) = 0.0$

and

Mean waiting time : $\bar{w}_i(n) = s_i \; (1+\bar{n}_i(n-1))$ [1]

Only the throughput expression explicitly requires evaluation of the normalization constant. Thus, if we could find an alternate method for computing the throughput of a station, we could evaluate all the performance measures without first evaluating the normalization constant.

The key idea is to apply Little's result to the entire queueing network [REIS78]. That is, let the "queueing system" be the entire network. Since we are considering a closed network, we know that the "average" number of jobs in the system is constrained to be n. The mean time a job

[1] Recall that this equation has an intuitive meaning: the mean waiting time a job experiences at station i is its own mean service time s_i plus the time it takes the station to perform its backlogged work $s_i\bar{n}_i(n-1)$.

spends in the system (before departing along the CPU-CPU path) is simply the sum of the time it spends at each of the individual service centers. But the average time a job spends at station i is equal to (mean number of visits to station i) x (mean waiting time per visit). The particular solution of the job flow balance equations chosen in step 1 above may also be interpreted as the mean number of visits per job to station i [DENN78b]. To emphasize this notion we will use the notation v_i rather than e_i. Now from Little's law

$$X_1(n) = n \Big/ \sum_{i=1}^{M} v_i \, \bar{w}_i(n)$$

(since in equilibium the arrival rate to a station must equal the throughput of that station). The throughputs of the other stations are computed from

$$X_i(n) = X_1(n) \, v_i \qquad i=2,3,\ldots,M.$$

Combining all these results, we formulate the following algorithm for computing the performance measures of a network with load independent stations [REIS78]:

(1) Solve the job flow balance equations for the v_i. (One particular solution is $v_1=1$ and $v_i=p_i$ for $i=2,3,\ldots,M$).

(2) Initialize $\bar{n}_i(0)=0.0$ for $i=1,2,\ldots,M$.

(3) Compute the performance measures of the network as follows:

```
for n:=1 to N do
begin
```

$$\bar{w}_i(n) = s_i\ (1+\bar{n}_i(n-1)) \qquad \text{for } i=1,2,\ldots,M$$

$$X_1(n) = n \Big/ \sum_{i=1}^{M} v_i\ \bar{w}_i(n)$$

$$X_i(n) = X_1(n)\ v_i \qquad \text{for } i=2,3,\ldots,M$$

$$U_i(n) = s_i\ X_i(n) \qquad \text{for } i=1,2,\ldots,M$$

$$\bar{n}_i(n) = X_i(n)\ \bar{w}_i(n) \qquad \text{for } i=1,2,\ldots,M$$

```
end
```

Thus, the simple performance measures of a central server model with load independent parameters can be computed using a surprisingly simple algorithm! Reiser and Lavenberg also described a recursive technique to evaluate the marginal queue length distribution : $p_i(n,N)$. This will be treated in the next section.

The amount of storage required by this algorithm is less than what appears evident. For example, one might declare an array for each of the performance measures with a structure resembling:

<u>array</u> [1..M,0..N] <u>of</u> <u>real</u>

In fact, the second dimension is superfluous, since we only require the previous value of n in any computation and not all values from 0 to n-1. Hence, arrays of the following structure suffice:

<u>array</u> [1..M] <u>of</u> <u>real</u>

With this observation it is easy to show that the mean value analysis algorithm for computing the performance measures of a queueing network requires less storage than the comparable normalization constant algorithm (since no memory is allocated for the normalization constant vector G). It is also a routine matter to show that the mean value analysis algorithm requires slightly fewer additions, multiplications, and considerably fewer divisions.

4.3 SUMMARY OF SINGLE CLASS MEAN VALUE ANALYSIS FORMULAS

We include this section for completeness and convenience. Since almost all the formulas used to calculate the performance measures of a si.igle class network have been derived in Chapter 2, we will simply list those formulas required by the mean value analysis approach here. Any formula not derived in Chapter 2 or this chapter is proved in Appendix A.

Marginal Queue Length Distribution

$$p_i(n,N) = \begin{cases} s_i X_i(N) p_i(n-1,N-1) & \text{for } n>0 \text{ load independent} \\ s_i(n) X_i(N) p_i(n-1,N-1) & \text{for } n>0 \text{ load dependent} \\ 1 & \text{for } n=0 \text{ and } N=0 \end{cases}$$

Denning and Buzen refer to this equation as the Prime Queue Recursion [DENN79].

Throughput

$$X_i(n) = \begin{cases} i=1 : & n \Big/ \sum_{j=1}^{M} v_j \, \bar{w}_j(n) \\ i \neq 1 : & X_1(n) \, v_i \end{cases}$$

Utilization

$$U_i(n) = \begin{cases} s_i \, X_i(n) & \text{load independent} \\ 1-p_i(0,n) & \text{load dependent} \end{cases}$$

Mean Queue Length

$$\bar{n}_i(n) = X_i(n) \, \bar{w}_i(n) \qquad \text{Little's formula}$$

Mean Waiting Time

$$\bar{w}_i(n) = \begin{cases} s_i & \text{IS station} \\ s_i \, (1+\bar{n}_i(n-1)) & \text{load independent} \\ \sum_{k=1}^{n} k \, s_i(k) \, p_i(k-1,n-1) & \text{load dependent} \end{cases}$$

The load dependent mean waiting time formula says that the mean waiting time of a station in a system with n jobs, is the average of the backlog that exists if each of the k jobs waiting requires time exactly equal to the mean time between completions (k=1,2,...,n). In agreement with principle 1. the weighting factor used in computing the average is the overall distribution at that station when n-1 jobs circulate throughout the system.

Note that the marginal queue length distribution must be calculated for any load dependent station since it is required in the evaluation of the mean waiting time. The computation of the queue length distribution can be accomplished using the following simple method. First, initialize $p_i(0,0)$ to 1.0:

```
for i:=1 to M do
    if station i is load dependent then pi(0)=1.0
```

Then in the load independent mean value analysis algorithm described in section 4.2, include the following code after the computation of the throughput ($X_i(n)$):

```
sum = 0.0                                  \
for np:=n downto 1 do                      |
begin                                      |
                                           |  for i=1,..,M
   pi(np) = si Xi(n) pi(np-1)              >  if station i
                                           |  is load
   sum = sum + pi(np)                      |  dependent
                                           |
end                                        |
pi(0) = 1.0 - sum                          /
```

Notice that the for loop counts backwards from n downto 1. This implementation saves storage since p need not be declared as a three dimensional array; instead all that it is required is:

```
p : array [1..M,0..N] of real
```

For a load dependent station the same algorithm applies, except that we use $s_i(np)$ instead of s_i.

4.4 MULTIPLE CLASS NETWORKS WITH LOAD INDEPENDENT PARAMETERS

The single class mean value analysis algorithm described in the section 4.2 generalizes directly into a multiple class algorithm as we will now demonstrate. (Table 4-1 summarizes the multiple class notation developed in Chapter 2.)

Table 4-1: Summary of Multiple Job Class Notation

$\underline{N}_R = (N_1, N_2, \ldots, N_R)$	the total number of jobs of each class.
v_{ir}	the mean number of visits each job of class r makes to station i.
$\bar{n}_{ir}(\underline{N}_R)$	the mean number of class r jobs queued at station i when $\underline{N}_R$ is the load of the system.
$\bar{w}_{ir}(\underline{N}_R)$	the mean waiting time for each class r job at station i when $\underline{N}_R$ is the load of the system.
$X_{ir}(\underline{N}_R)$	the throughput of class r jobs from station i when $\underline{N}_R$ is the load of the system.
s_{ir}	the mean time between completions of class r jobs from station i.

Again we restrict our attention to central server networks with M load independent FCFS stations. Let there be R distinct job classes where jobs are not permitted to switch class. The mean service time per request of station i by a job in class r is s_{ir}. Then the Reiser-Lavenberg algorithm is:

(1) Solve the job flow balance equations for the visit ratios v_{ir}. (One particular solution is $v_{1r}=1$ for $r=1,2,...,R$ and $v_{ir}=p_{(1,r)(i,r)}$ for $i=2,3,...,M$ and $r=1,2,...,R$).

(2) Initialize $\bar{n}_i(\underline{0}_R) = 0$ for $i=1,2,...,M$.

(3) Compute the performance measures of the network as follows:

<u>for</u> n_1:=0 <u>to</u> N_1 <u>do</u>
  <u>for</u> n_2:=0 <u>to</u> N_2 <u>do</u>
    .
    .
      <u>for</u> n_R:=0 <u>to</u> N_R <u>do</u>
      <u>begin</u>

$$\underline{n}_R = (n_1, n_2, ..., n_R)$$

$$\bar{w}_{ir}(\underline{n}_R) = s_{ir}(1+\bar{n}_i(\underline{n}_R-\underline{1}_r)) \quad \left\{ \text{for } 1 \leqq i \leqq M \text{ and } 1 \leqq r \leqq R \right.$$

$$X_{1r}(\underline{n}_R) = n_r \Big/ \sum_{i=1}^{M} v_{ir}\, \bar{w}_{ir}(\underline{n}_R) \quad 1 \leqq r \leqq R$$

$$X_{ir}(\underline{n}_R) = X_{1r}(\underline{n}_R)\, v_{ir} \quad \left\{ \text{for } 2 \leqq i \leqq M \text{ and } 1 \leqq r \leqq R \right.$$

$$\left.\begin{aligned} U_{ir}(\underline{n}_R) &= s_{ir}\, X_{ir}(\underline{n}_R) \\ \bar{n}_{ir}(\underline{n}_R) &= X_{ir}(\underline{n}_R)\, \bar{w}_{ir}(\underline{n}_R) \end{aligned}\right\} \text{for } 1 \leqq i \leqq M \text{ and } 1 \leqq r \leqq R$$

$$\bar{n}_i(\underline{n}_R) = \sum_{r=1}^{R} \bar{n}_{ir}(\underline{n}_R) \quad \text{for } 1 \leqq i \leqq M$$

      <u>end</u>

For simplicity we have assumed that every class visits each

station. Since we are considering FCFS stations, the BCMP theorem requires that $s_{i1}=s_{i2}=\cdots s_{iR}$. Also, we assume that reference to elements with negative indices yields the value zero.

The R nested for loops in the above algorithm can be replaced by the repeat-until construct discussed in section 3.4.

The implementation of this algorithm requires a set of arrays to store intermediate results. In particular, the following arrays would be required

$\bar{n}_{ir}, U_{ir}, v_{ir}, \bar{w}_{ir}, X_{ir}$: <u>array</u> [1..M,1..R] <u>of</u> <u>real</u>;

and

$\bar{n}_i$: <u>array</u> [1..M] <u>of</u>
<u>array</u> [0..N_1,0..N_2,...,0..N_R] <u>of</u> <u>real</u>;

The arrays $\bar{n}_{ir}$, U_{ir} , v_{ir} , $\bar{w}_{ir}$, and X_{ir} each require M*R storage locations. The problem with this algorithm as written is the storage space required for the array $\bar{n}_i$, i.e.,

$$M * \prod_{r=1}^{R} (N_r + 1) \quad \text{locations.}$$

For a moderately small network containing M=5 stations, and R=4 job classes with $N_1=N_2=N_3=N_4=19$ jobs/class, the storage requirement for $\bar{n}_i$ alone would be 800,000 words of memory!

Fortunately, there is a way to reduce this storage requirement considerably. We demonstrate the idea by a simple example of a two class network with $N_1=2$ and $N_2=2$. The states generated by the nested for loops are listed below in order of their generation.

n_1	n_2
0	0
0	1
0	2
1	0
1	1
1	2
2	0
2	1
$N_1 = 2$	$2 = N_2$

Notice that as soon as state (1,0) is computed using state (0,0), state (0,0) is no longer required in the generation of any other state. Hence, the value computed for state (1,0) can overwrite the memory location used to contain the value of state (0,0). So, for this example we see that only N_2+1 storage locations are required. The same idea easily extends to more than two classes, the storage required for one station being

$$\prod_{r=2}^{R} (N_r+1)$$

If we call this quantity "maxsize", then the recommended

array declaration becomes:

```
n̄i : array [1..M] of
        array [0..maxsize-1] of real;
```

This declaration avoids the compile time restriction of having to know how many job classes will be present in the network (i.e., the value of R at execution time).

Also, notice to take full advantage of this storage allocation scheme, classes should be ordered so that class 1 contains the most number of jobs. A simple index function can be used to compute the proper location of any state in the array $\bar{n}_i$.

The total storage requirement for the array $\bar{n}_i$ is

$$M * \prod_{r=2}^{R} (N_r+1)$$

This compares favorably with the storage required by the normalization constant approach using two G vectors which occupy

$$2 * \prod_{r=1}^{R} (N_r+1) \quad \text{memory locations}$$

4.5 SUMMARY OF MULTIPLE CLASS MEAN VALUE ANALYSIS FORMULAS

In this section we collect together all the required multiple class mean value analysis formulas. Again, those formulas not derived in Chapter 2 are proved in Appendix A.

Marginal Queue Length Distribution

$$p_i(\underline{n}_R, \underline{N}_R) = \begin{cases} \sum_{r=1}^{R} s_{ir}\, X_{ir}(\underline{N}_R)\, p_i(\underline{n}_R - \underline{1}_r, \underline{N}_R - \underline{1}_r) & \text{load independent} \\ \sum_{r=1}^{R} s_{ir}(n_{ir})\, X_{ir}(\underline{N}_R)\, p_i(\underline{n}_R - \underline{1}_r, \underline{N}_R - \underline{1}_r) & \text{load dependent} \end{cases}$$

Throughput

$$X_{ir}(\underline{n}_R) = \begin{cases} i=1: \; n_r \Big/ \sum_{j=1}^{M} v_{jr}\, \bar{w}_{jr}(\underline{n}_R) \\ i\neq 1: \; X_{1r}(\underline{n}_R)\, v_{ir} \end{cases}$$

Utilization

$$U_{ir}(\underline{n}_R) = \begin{cases} s_{ir}\, X_{ir}(\underline{n}_R) & \text{load independent} \\ 1 - p_i(0, \underline{n}_R) & \text{load dependent} \end{cases}$$

Mean Queue Length

$$\bar{n}_{ir}(\underline{n}_R) = X_{ir}(\underline{n}_R)\, \bar{w}_{ir}(\underline{n}_R)$$ Little's formula

$$\bar{n}_i(\underline{n}_R) = \sum_{r=1}^{R} \bar{n}_{ir}(\underline{n}_R)$$ Station mean queue length independent of class

Mean Waiting Time

$$\bar{w}_{ir}(\underline{n}_R) = \begin{cases} s_{ir} & \text{IS station} \\ s_{ir}(1+\bar{n}_i(\underline{n}_R-\underline{1}_r)) & \text{load independent} \\ s_{ir} + \sum_{s=1}^{R} s_{is}\ \bar{n}_{is}(\underline{n}_R-\underline{1}_r) & \text{load independent FCFS (identical to above if } s_{i1}=s_{i2}=\cdots s_{iR}) \\ \sum_{\underline{k}_R=\underline{0}_R}^{\underline{n}_R} \left(\sum_{r=1}^{R} k_r\ s_{ir}(k_r)\right) p_i(\underline{k}_R-\underline{1}_r,\underline{n}_R-\underline{1}_r) & \text{load dependent} \end{cases}$$

The second expression for the load independent FCFS mean waiting time is exact if $s_{i1}=s_{i2}=\cdots=s_{iR}$ (as required by the BCMP theorem). When the mean service times are not equal, the expression is only an approximation [BARD79].

4.6 EXTENSIONS TO SWITCHING BETWEEN CLASSES

Extending mean value analysis to account for jobs switching class membership is straightforward. (The same is true for the normalization constant approach -- cf. Chapter 2.)

We partition the original R classes into $U \leq R$ equivalence classes (the sets denoted by EC_1, EC_2, ..., EC_U) such that jobs do not change equivalence class. Since jobs can switch class membership, the number of jobs in each class is not a constant -- but the number of jobs in each equivalence class does remain constant. If we let N'_r denote the initial number of jobs in class r (we used N_r previously), then

$$N_q = \sum_{r \in EC_q} N'_r \qquad \text{for } q=1,2,\ldots,U$$

denotes the number of jobs belonging to "equivalence class q", (EC_q); the distribution of jobs over equivalence classes is denoted by the vector $\underline{N}_U = (N_1, N_2, \ldots, N_U)$. See section 2.4 for further details on the partitioning of jobs into equivalence classes.

We will now show how to compute performance measures by equivalence class from which the performance measures by job class are easily obtained. Let v^*_{iq} be the visit ratio of equivalence class q jobs to station i. Then,

$$v^*_{iq} = \frac{\sum_{r \in EC_q} v_{ir}}{\sum_{r \in EC_q} v_{1r}} \qquad \text{for } q=1,2,\ldots,U$$

Also, define this ratio:

$$\alpha_{ir} = \frac{v_{ir}}{\sum_{s \in EC_q} v_{is}} \qquad \text{for } r=1,2,\ldots,R$$

which represents the proportion of class r jobs (from those in equivalence class q) which visit station i. Let $\overline{w}^*_{iq}$ represent the mean waiting time of equivalence class q jobs at station i; then

$$\bar{w}^*_{iq}(\underline{n}_U) = s_{iq}(1+\bar{n}^*_i(\underline{n}_U-\underline{1}_q)) \qquad \text{for } q=1,2,\ldots,U$$

where $\bar{n}^*_i(\underline{n}_U)$ represents the total mean queue length at station i; it is simply the sum of the mean queue lengths by equivalence class at station i. If X^*_{iq} denotes the throughput of equivalence class q jobs from station i we have:

$$X^*_{1q}(\underline{n}_U) = n_q \Big/ \sum_{i=1}^{M} v^*_{iq}\ \bar{w}^*_{iq}(\underline{n}_U) \qquad \text{for } q=1,2,\ldots,U$$

Then for stations other than the first:

$$X^*_{iq}(\underline{n}_U) = X^*_{1q}(\underline{n}_U)\ v^*_{iq} \qquad \text{for } q=1,2,\ldots,U$$

This throughput is the sum of the throughputs of each class belonging to equivalence class q; thus the throughput by job class can be computed as:

$$X_{ir}(\underline{n}_U) = \alpha_{ir}\ X^*_{iq}(\underline{n}_U) \qquad q \text{ such that } r\in EC_q \text{ and } 1\leq r\leq R$$

Hence, we are able to recover per class throughputs (X_{ir}) from per equivalence class throughputs (X^*_{iq}).

The mean queue length by class is calculated from Little's formula:

$$\bar{n}_{ir}(\underline{n}_U) = X_{ir}(\underline{n}_U)\ \bar{w}_{ir}(\underline{n}_U) \qquad \text{for } r=1,2,\ldots,R$$

where the mean waiting time is given by

$$\bar{w}_{ir}(\underline{n}_U) = s_{ir}(1+\bar{n}^*_i(\underline{n}_U-\underline{1}_q)) \qquad \text{for } r=1,2,\ldots,R$$

The algorithm for the multiple class central server network with M load independent FCFS stations now becomes:

(1) Solve the job flow balance equations for the visit

ratios v_{ir}.

(2) Determine the equivalence classes EC_1, $EC_2, \ldots, EC_U$.

(3) Compute v^*_{iq} for $i=1,2,\ldots,M$ and $q=1,2,\ldots,U$.

(4) Compute α_{ir} for $i=1,2,\ldots,M$ and $r=1,2,\ldots,R$

(5) Initialize $\bar{n}^*_i(\underline{0}_U)=0$ for $i=1,2,\ldots,M$ and $r=1,2,\ldots,R$

(6) Compute the performance measures as follows:

```
n:=zero;                    (*initialize vector n to zero*)
n[1]:=1;
repeat
```

$$\underline{n}_U = (n_1, n_2, \ldots, n_U)$$

$$\bar{w}^*_{iq}(\underline{n}_U) = s_{iq}(1+\bar{n}^*_i(\underline{n}_U-\underline{1}_q)) \qquad \text{for } 1 \leq i \leq M \text{ and } 1 \leq q \leq U$$

$$X^*_{1q}(\underline{n}_U) = n_q \Big/ \sum_{i=1}^{M} v^*_{iq}\bar{w}^*_{iq}(\underline{n}_U) \qquad \text{for } 1 \leq q \leq U$$

$$X^*_{iq}(\underline{n}_U) = X^*_{1q}(\underline{n}_U)\ v^*_{iq} \qquad \text{for } 2 \leq i \leq M \text{ and } 1 \leq q \leq U$$

$$U^*_{iq}(\underline{n}_U) = s_{iq}\ X^*_{iq}(\underline{n}_U)$$

$$\bar{n}^*_{iq}(\underline{n}_U) = X^*_{iq}(\underline{n}_U)\ \bar{w}^*_{iq}(\underline{n}_U) \qquad \text{for } 1 \leq i \leq M \text{ and } 1 \leq q \leq U$$

$$\bar{n}^*_i(\underline{n}_U) = \sum_{q=1}^{U} \bar{n}^*_{iq}(\underline{n}_U) \qquad \text{for } 1 \leq i \leq M$$

```
until ¬sum(n,N,U);
```

Notice that this algorithm is isomorphic to the multiple class algorithm of section 4.4. At any step in the algorithm the per class performance measures can be computed as follows:

$$\bar{w}_{ir}(\underline{n}_U) = s_{ir}(1+\bar{n}^*_i(\underline{n}_U-\underline{1}_q))$$

$$X_{ir}(\underline{n}_U) = \alpha_{ir}\; X^*_{iq}(\underline{n}_U)$$

$$U_{ir}(\underline{n}_U) = s_{ir}\; X_{ir}(\underline{n}_U)$$

$$\bar{n}_{ir}(\underline{n}_U) = X_{ir}(\underline{n}_U)\; \bar{w}_{ir}(\underline{n}_U)$$

Observe that the mean waiting time per class is equal to the mean waiting time per equivalence class.

4.7 CONCLUSION

In this chapter we have shown that mean value analysis is an attractive alternative to the normalization constant approach of Chapter 2. Among the advantages of this approach we can mention its simplicity (which makes it very easy to program), and its numerical stability since no big numbers are now generated. Hence, we avoid scaling (cf. section 3.3) which was necessary in the normalization constant approach to control overflow and/or underflow [REIS75,REIS78,BRUE78]. The mean value analysis algorithms are slightly more computationally efficient than the corresponding normalization constant algorithms. Also, the storage requirement of the mean value analysis algorithms is comparable to that used by the normalization constant algorithms -- provided that the implementation strategies of this chapter are followed.

CHAPTER 5

SUMMARY

5.1 SUMMARY AND CONCLUSIONS

This book has treated the algorithms for evaluating multi-class queueing network models of computer systems. The contributions of this book are three: First (Chapter 2): we presented a detailed derivation of the computation of performance measures in single and multiple job class queueing network models. We showed that the multiple class computational algorithms are extensions of the single class algorithms. These extensions can be expressed in a convenient vector notation that emphasizes the similarity between the single and multiple class algorithms. For example, the single class load independent algorithm for computing the normalization constant is:

$$g_m(n\ ) = g_{m-1}(n\ ) + \quad Y_m\ g_m(n\ \ - 1\ ).$$

Its multiple class analogue is a natural vector extension:

$$g_m(\underline{n}_R) = g_{m-1}(\underline{n}_R) + \sum_{r=1}^{R} Y_{mr}\ g_m(\underline{n}_R - \underline{1}_r)$$

We also demonstrated that class switching actually reduces the computational complexity of a queueing network model.

Second (Chapter 3): we investigated certain practical problems involved with implementing the algorithms of Chapter 2 as a software package. We showed a simple scaling rule that controls overflow and underflow in the algorithms for calculating the normalization constant. We also showed how to simplify networks with multiple infinite server (IS) stations: all the IS stations can be replaced by a single IS station at which special job classes simulate the original IS stations.

Third (Chapter 4): we presented Reiser and Lavenberg's mean value analysis of queueing network models. We showed that their method is an alternative solution to the normalization constant method for product form queueing networks. We showed that the storage requirement of the mean value analysis algorithms is comparable to that of the corresponding normalization constant algorithm, but that the mean value analysis algorithms are considerably easier to program and understand.

5.2 RECOMMENDATIONS

The performance measures of a queueing network can be computed either by using the normalization constant (NC) method of Chapter 2 or the mean value analysis (MVA) method of Chapter 4. Each has advantages and limitations. The computational speed of these methods is not an issue because the algorithms execute quite rapidly. The choice between the two methods must be based on other considerations, such as the set of performance measures to be calculated, the storage requirements of the requisite computational algorithms, and the numerical accuracy (stability) of these algorithms when applied to the specific problem.

For example, if only the system throughput function of the network is required, the NC method is preferred because we can calculate this function simply once the normalization constant vector has been computed. In contrast, the MVA method requires that we also compute the mean waiting time, throughput, and mean queue length of each station even though we are not interested in these quantities at all. In this special case the NC method requires about half as much storage as the MVA method.

The storage requirement of the NC method is always less than or equal to that of the MVA algorithms. Recall, though, that if the guidelines of Chapter 4 are followed, the MVA algorithms will not require significantly more storage than the NC algorithms. When all performance metrics are desired, therefore, neither speed nor storage requirement should be the deciding factor.

Overflow and underflow problems may arise in the computation of the normalization constant. The scaling rule we presented in Chapter 3, while not necessarily the best [cf. for example, REIS75 or CHAN79], alleviates (or at least postpones) underflow or overflow. No scaling is necessary in the MVA algorithms since no big numbers are generated.

The most serious drawback of the MVA approach is the accumulative effect of round-off errors which propogate forward in the iterations from one network load to the next. These errors can be especially serious in the recursive computation of queue length distributions, especially when there is a wide disparity among speeds of stations in the network. Estimates of probabilities outside the range [0,1] may occur [CHAN79]. No similar problem is known for the NC method. Therefore, it is recommended to use the NC approach if and when queue length distributions are required.

Bard [BARD79] and Reiser [REIS79] have suggested fast approximations based on mean value analysis. Because the product form queueing network model is itself an approximation to the real system, we do not know whether the

Bard-Reiser approximations improve the estimates of standard performance metrics. They are, however, very simple to program. Recall that the iteration scheme of MVA begins with estimates of the mean queue lengths for network load $\underline{n}_R - \underline{1}_r$ and computes new estimates of these (and other quantities) for load $\underline{n}_R$. An approximation suggested by P. Schweitzer of the University of Rochester and J. Zahorjan of the University of Toronto is that the mean queue length in class r for load $\underline{n}_R - \underline{1}_r$ is linearly scaled down from the mean queue length in class r for load $\underline{n}_R$:

$$\bar{n}_{ir}(\underline{n}_R - \underline{1}_r) = \frac{n_r - 1}{n_r} \bar{n}_{ir}(\underline{n}_R)$$

On introducing this approximation, one can drop the explicit dependence of the metrics on load $\underline{n}_R$ and simplify the equations on page 144 of Chapter 4 to these:

$$\bar{w}_{ir} = s_{ir}(1+\bar{n}_i) \qquad \left\{ \begin{array}{l} \text{for } 1 \leqq i \leqq M \\ \text{and} \\ 1 \leqq r \leqq R \end{array} \right.$$

$$X_{1r} = n_r \Big/ \sum_{i=1}^{M} v_{ir} \bar{w}_{ir} \qquad 1 \leqq r \leqq R$$

$$X_{ir} = X_{1r} v_{ir} \qquad \left\{ \begin{array}{l} \text{for } 2 \leqq i \leqq M \\ \text{and} \\ 1 \leqq r \leqq R \end{array} \right.$$

$$U_{ir} = s_{ir} X_{ir}$$

$$\bar{n}_{ir} = X_{ir} \bar{w}_{ir}$$

$$\left\{ \text{for } 1 \le i \le M \text{ and } 1 \le r \le R \right.$$

$$\bar{n}_i = \sum_{r=1}^{R} \bar{n}_{ir} \qquad \text{for } 1 \le i \le M$$

These equations can be solved by choosing an initial guess for the $\bar{n}_{ir}$, and refining the guess iteratively until successive sets of $\bar{n}_{ir}$ do not differ by much. The initial guess should satisfy the requirement that

$$\sum_i \bar{n}_{ir} = N_r$$

for all r; a simple such guess is $\bar{n}_{ir} = N_r/M$. The above iteration always converges [BARD79] and usually does so rapidly. The basic scheme can be modified to yield approximate solutions for queues having priorities [BARD79,BALB79b].

All the algorithms discussed in this book have been implemented in the PNET package, a proprietary software system that operates at Purdue University and the University of Minnesota. These and similar algorithms have been implemented in commercially available queueing network solvers. All the algorithms have been extensively tested.

LIST OF REFERENCES

[BALB79a] Balbo, G.; and Denning, P.J. "Homogeneous Approximations of General Queueing Networks," Proc. 4-th International Symposium on Modelling and Performance Evaluation of Computer Systems, Vienna, Austria, (February 1979).

[BALB79b] Balbo, G. "Approximate Solutions of Queueing Network Models of Computer Systems," Ph.D. Thesis, Computer Sciences Department, Purdue University, (1979), in progress.

[BARD79] Bard, Y. "Some Extensions to Multiclass Queueing Network Analysis," Proc. 4-th International Symposium on Modelling and Performance Evaluation of Computer Systems, Vienna, Austria, (February 1979).

[BASK72] Baskett, F.; and Palacios, F.G. "Processor Sharing in a Central Server Queueing Model of Multiprogramming with Applications," Proc. Sixth Annual Princeton Conference on Information Sciences and Systems, Princeton University, Princeton, New Jersey, (March 1972), 598-603.

[BASK73] Baskett, F.; and Muntz, R.R. "Networks of Queues," Proc. of the 7th Annual Princeton Conference on Information Sciences and Systems, Princeton University, Princeton, New Jersey, (March 1973), 428-434.

[BASK75] Baskett, F.; Chandy, K.M.; Muntz, R.R.; and Palacios, F. "Open, Closed, and Mixed Networks of Queues with Different Classes of Customers," J. ACM 22, 2 (April 1975), 248-260.

[BRUE78] Bruell, S.C. "On Single and Multiple Job Class Queueing Network Models of Computing Systems," Ph.D. Thesis, Computer Sciences Department, Purdue University, (December 1978).

[BUZE71a] Buzen, J.P. "Analysis of System Bottlenecks Using a Queueing Network Model," Proc. ACM Workshop on System Performance Evaluation, Harvard University, (March 1971), 82-102.

[BUZE71b] Buzen, J.P. "Queueing Network Models of Multiprogramming", Ph.D. Thesis, Division of Engineering and Applied Science, Harvard University, Cambridge, Massachusetts, (August 1971).

[BUZE73] Buzen, J.P. "Computational Algorithms for Closed Queueing Networks with Exponential Servers," Comm. ACM 16, 9 (September 1973), 527-531.

[BUZE76] Buzen, J.P. "Operational Analysis: The Key to the New Generation of Performance Prediction Tools," Proc. IEEE Compcon, (September 1976).

[BUZE76a] Buzen, J.P. "Operational Analysis: An Alternative to Stochastic Modeling," Technical Report, BGS Systems, Inc., Box 128, Lincoln, MA 01773, (October 1976).

[BUZE76c] Buzen, J.P. "Fundamental Laws of Computer System Performance," Proc. International Symp. on Computer Modeling, Measurement and Evaluation, ACM-SIGMETRICS and IFIP WG7.3, (March 1976), 200-210.

[BUZE78b] Buzen, J.P.; et al. "BEST/1 – Design of a Tool for Computer System Capacity Planning," Proc. 1978 AFIPS National Computer Conference 47, AFIPS Press, Montvale N.J., 447-455.

[CHAN72] Chandy, K.M. "The Analysis and Solutions for General Queueing Networks," Proc. of the 6th Annual Princeton Conference on Information Sciences and Systems, Princeton University, (March 1972), 224-228.

[CHAN75a] Chandy, K.M.; Herzog, U.; and Woo, L. "Parametric Analysis of Queueing Network Models," IBM J. of R. & D. 19, 1 (January 1975), 36-42.

[CHAN75b] Chandy, K.M.; Herzog, U.; and Woo, L. "Approximate Analysis of General Queueing Networks," IBM J. of R. & D. 19, 1, (January 1975), 43-49.

[CHAN79] Chandy, K.M; and Sauer, C.H. "Computational Algorithms for Product Form Queueing Networks," IBM Tech Report RC7950.

[COOP72] Cooper, R.B. "Introduction to Queueing Theory", The MacMillan Company, New York, (1972).

[COUR71] Courtois, P.J. "On the Near-Complete-Decomposability of Networks of Queues and of Stochastic Models of Multiprogrammed Computer Systems," Sci. Rpt. CMU-CS-71-11, Carnegie-Mellon University, (November 1971).

[COUR75] Courtois, P.J. "Decomposability, Instabilities, and Saturation in Multiprogramming Systems," Comm. ACM 18, 7 (July 1975), 371-376.

[COUR77] Courtois, P.J. "Decomposability: Queueing and Computer System Applications," Academic Press, (1977).

[DENN78] Denning, P.J. "Optimal Multiprogrammed Memory Management," in Advances in Programming Methodology III (R. Yeh and K. M. Chandy, Eds.) Prentice-Hall, (1978).

[DENN78b] Denning, P.J. ; and Buzen, J.P. "The Operational Analysis of Queueing Network Models," Computing Surveys 10, 3 (Sept. 1978).

[DENN79] Denning, P.J.; and Buzen, J.P. "An Operational Treatment of Queueing Distributions and Mean Value Analysis," August 1979, Tech Report CSD-TR 309, Purdue University.

[FERD71] Ferdinand, A.E. "An Analysis of the Machine Interference Model," IBM Systems Journal 10, 2 (1971), 129-142.

[GAVE71] Gaver, D.P. "Analysis of Remote Terminal Backlogs under Heavy Demand Conditions," J. ACM 18, 3 (1971), 405-415.

[GAVE73] Gaver, D.P.; and Shedler, G.S. "Processor Utilization in Multiprogramming Systems via Diffusion Approximations," Oper. Res. 21 , (1973), 569-576.

[GELE75] Gelenbe, E. "On Approximate Computer System Models," J. ACM 22, 2 (April 1975), 261-269.

[GORD67a] Gordon, W.J.; and Newell, G.J. "Closed Queueing Systems with Exponential Servers," Operations Research 15, (March 1967), 254-265.

[GORD67b] Gordon, W.J.; and Newell, G.J. "Acknowledgment," Operations Research, (December 1967), 1182.

[GRAH78] Graham, G.S. ACM Computing Surveys 10, 3, "Special Issue: Queueing Network Models of Computer System Performance," Guest Editor: G. S. Graham, (September 1978).

[JACK57] Jackson, J.R. "Networks of Waiting Lines," Operations Research 5 (1957), 518-521.

[JACK63] Jackson, J.R. "Jobshop-like Queueing Systems," Management Science 10, (October 1963), 131-142.

[KLEI75] Kleinrock, L. Queueing Systems, Volume 1: Theory. Wiley-Interscience, New York, (1975).

[KOBA74a] Kobayashi, H. "Applications of the Diffusion Approximation to Queueing Networks: Part I - Equilibrium Queue Distributions," J. ACM 21,2 (1974), 316-328.

[KOBA74b] Kobayashi, H. "Applications of the Diffusion Approximation to Queueing Networks: Part II - Transient Queue Distributions," J. ACM 21, 3 (1974), 459-468.

[KOBA75] Kobayashi, H. "System Design and Performance Analysis Using Analytic Models," IBM Research Report RA75, IBM Thomas J. Watson Research Center, Yorktown Heights, N.Y., (December 1975).

[LAVE79] Lavenberg, S.S.; and Reiser, M. "Stationary State Probabilities at Arrival Instants for Closed Queueing Networks with Multiple Types of Customers," IBM Research Report RC 7592, (April 1979).

[LITT61] Little, J.D.C. "A Proof of the Queueing Formula L=λW," Operations Research 9, (1961), 383387.

[MARI79] Marie, R."An Approximate Analytical Method for General Queueing Networks," Rapport IRISA, Universite' de Rennes, France, (1979).

[MUNT72] Muntz, R.R. "Poisson Departure Processes and Queueing Networks," IBM Research Report RC 4145, (December 1972).

[MUNT75] Muntz, R.R.. "Analytic Modeling of Interactive Systems," Proc. IEEE: Special Issue on Interactive Computer Systems, (June 1975), 946-953.

[MUNT78] Muntz, R.R. "Queueing Networks: A Critique of the State of the Art and Directions for the Future," Computing Surveys, 3, 10 (September 1978), 353-359.

[POSN68] Posner, M. and B. Bernholtz. "Closed Finite Queueing Networks with Time Lags and with Serveral Classes of Units," Operations Research 16, (1968), 977-985.

[REIS74] Reiser, M.; and Kobayashi, H. "Accuracy of the Diffusion Approximation for some Queueing Systems," IBM J of R. & D. 18, (March 1974), 114-124.

[REIS75] Reiser, M.; and Kobayashi, H. "Queueing Networks with Multiple Closed Chains: Theory and Computational Algorithms," IBM J. of R. & D. 19, 3 (May 1975), 283-294.

[REIS78] Reiser, M.; and Lavenberg, S.S. "Mean Value Analysis of Closed Multichain Queueing Networks," IBM T. J. Watson Research Center, RC 7023, (March 1978).

[REIS79] Reiser, M. "A Queueing Network Analysis of Computer Communication Networks with Window Flow Control," IEEE Transactions on Communications, (August 1979), 1199-1209.

[SAUE75] Sauer, C.H. "Configuration of Computing Systems: an Approach Using Queueing Network Models," Ph.D. Thesis, The University of Texas at Austin, (May 1975).

[SCHE67] Scherr, A.L. "An Analysis of Time Shared Computer Systems," MIT Press, (1967).

[SEVC77] Sevcik, K.C.; Levy, A.I.; Tripathi, S.L.; and Zahorjan, J.L. "Improving Approximations of Aggregated Queueing Network Subsystems," Proc. International Symposium on Computer Performance

Modeling, Measurement, and Evaluation, North-Holland Publishing Co., (August 1977), 1-22.

[SEVC79] Sevcik, K.C.; and Mitrani, I. "The Distribution of Queueing Network States at Input and Output Instants," Proc. 4-th International Symposium on Modelling and Performance Evaluation of Computer Systems, Vienna, Austria, (February 1979).

[SHUM76] Shum, A.W.C. "Queueing Models for Computer Systems with General Service Time Distributions," Ph.D. Thesis, Division Engrg. and Applied Physics, Harvard Univ., Cambridge, MA 02138, (December 1976).

[SHUM77] Shum, A.W.C.; and Buzen, J.P. "The EPF Technique: a Method for Obtaining Approximate Solutions to Closed Queueing Networks with General Service Times," Proc. 3rd International Symposium on Modeling and Performance Evaluation of Computer Systems, North-Holland Publishing Co., (October 1977).

[SIMO79] Simon, Richard T. "Modeling of Virtual Memory Systems," Ph.D Thesis, Purdue University, in progress.

[STEW77] Stewart, W.A. "A New Approach to the Numerical Analysis of Markovian Models,"Proc. International Symposium on Computer Performance Modeling, Measurement, and Evaluation, North-Holland Publishing Co., (August 1977), 279-295.

[WONG75] Wong, J.W. "Queueing Network Models for Computer Systems," Ph.D. Thesis, Computer Science Department, University of California at Los Angeles, (October 1975).

APPENDIX

PROOFS OF EQUATIONS IN CHAPTERS 2 AND 4

INTRODUCTION

This appendix contains proofs for the equations not proved in the body of Chapters 2 and 4. The notation is the same as that employed throughout those chapters and will therefore not be repeated here.

Eq. 2.2:

$$g_m(n) = g_{m-1}(n) + Y_m \, g_m(n-1)$$

Proof:

$$g_m(n) = \sum_{k=0}^{n} f_m(k) \, g_{m-1}(n-k)$$

$$= g_{m-1}(n) + \sum_{k=1}^{n} f_m(k)\; g_{m-1}(n-k)$$

$$= g_{m-1}(n) + \sum_{k=0}^{n-1} f_m(k+1)\; g_{m-1}(n-1-k)$$

$$= g_{m-1}(n) + Y_m \sum_{k=0}^{n-1} f_m(k)\; g_{m-1}(n-1-k)$$

$$= g_{m-1}(n) + Y_m\; g_m(n-1)$$

<u>Proof that $g_M^M(n) = g_{M-1}(n)$ for n=0,1,...,N:</u>

$$g_M^M(n) = \sum_{\substack{\underline{n}\in S(N,M)\\ n_M=N-n}} \prod_{\substack{j=1\\ j\neq M}}^{M} f_j(n_j)$$

$$= \sum_{\underline{n}\in S(n,M-1)} \prod_{j=1}^{M-1} f_j(n_j)$$

$$= g_{M-1}(n)$$

<u>Eq. 2.5</u>:

$$g_M^i(n) = G(n) - Y_i\; G(n-1)$$

<u>Proof</u>:

$$g_M^i(n) = G(n) - \sum_{k=1}^{n} f_i(k)\, g_M^i(n-k)$$

$$= G(n) - \sum_{k=0}^{n-1} f_i(k+1)\, g_M^i(n-1-k)$$

$$= G(n) - Y_i \sum_{k=0}^{n-1} f_i(k)\, g_M^i(n-1-k)$$

$$= G(n) - Y_i\, G(n-1)$$

<u>Eq. 2.6</u>:

$$g_M^i(n)=G(n) - \frac{Y_i}{a_i} \left[G(n-1) + \sum_{k=0}^{a_i-2} \binom{a_i-k-1}{k+1} f_i(k)\, g_M^i(n-1-k)\right]$$

<u>Proof</u>:

$$g_M^i(n) = G(n) - \sum_{k=1}^{n} f_i(k)\, g_M^i(n-k)$$

$$= G(n) - Y_i \sum_{k=1}^{n} d_i(k)\, f_i'(k-1)\, g_M^i(n-k) \quad (f_i'(k)=Y_i^{\,k})$$

$$= G(n) - Y_i \sum_{k=0}^{n-1} d_i(k+1)\, f_i'(k)\, g_M^i(n-1-k)$$

$$= G(n) - Y_i [\sum_{k=0}^{a_i-1} \frac{1}{k+1} f_i(k) \; g_M^i(n-1-k) +$$

$$\sum_{k=a_i}^{n-1} \frac{1}{a_i} f_i(k) \; g_M^i(n-1-k)]$$

$$= G(n) - Y_i [\frac{G(n-1)}{a_i} + \sum_{k=0}^{a_i-2} \left(\frac{1}{k+1} - \frac{1}{a_i} \right) f_i(k) g_M^i(n-1-k)]$$

$$= G(n) - \frac{Y_i}{a_i} [G(n-1) + \sum_{k=0}^{a_i-2} \left(\frac{a_i-k-1}{k+1} \right) f_i(k) \; g_M^i(n-1-k)]$$

<u>Load independent mean queue length</u>:

$$\bar{n}_i(N) = \sum_{n=1}^{N} Y_i^{\,n} \frac{G(N-n)}{G(N)}$$

<u>Proof</u>: Define the tail of the queue length distribution to be

$$Q_i(n) = \sum_{\substack{\underline{n} \in S(N,M) \\ n_i \geqq n}} P(n_1, n_2, \ldots, n_M)$$

$$= \sum_{\substack{\underline{n} \in S(N,M) \\ n_i \geqq n}} \frac{1}{G(N)} \prod_{j=1}^{M} f_j(n_j)$$

$$= \frac{f_i(n)}{G(N)} \sum_{\underline{n} \in S(N-n,M)} \prod_{j=1}^{M} f_j(n_j)$$

$$= \frac{Y_i^{\,n}}{G(N)} \; G(N-n)$$

But

$$\bar{n}_i(N) = \sum_{n=1}^{N} Q_i(n). \qquad \text{QED}$$

<u>Proof that $\bar{n}_i(N) = U_i(N)(1+\bar{n}_i(N-1))$</u>:

$$\bar{n}_i(N) = \sum_{n=1}^{N} Y_i^{\,n} \frac{G(N-n)}{G(N)}$$

$$= Y_i \frac{G(N-1)}{G(N)} + \sum_{n=2}^{N} Y_i^{\,n} \frac{G(N-n)}{G(N)}$$

$$= U_i(N) + \frac{Y_i}{G(N)} \sum_{n=1}^{N-1} Y_i^{\,n} \; G(N-1-n)$$

$$= U_i(N) + U_i(N) \sum_{n=1}^{N-1} Y_i^{\,n} \frac{G(N-1-n)}{G(N-1)}$$

$$= U_i(N)(1+\bar{n}_i(N-1))$$

Lemma 1: If station m is load independent and has a FCFS, PS, or LCFS service discipline, then

$$f_m(\underline{n}_R) = \sum_{r=1}^{R} Y_{mr}\, f_m(\underline{n}_R - \underline{1}_r) \ .$$

Proof:

$$f_m(\underline{n}_R) = \frac{(n_{m1}+n_{m2}+\cdots+n_{mR})!}{n_{m1}!\,n_{m2}!\cdots n_{mR}!} Y_{m1}^{n_{m1}} Y_{m2}^{n_{m2}} \cdots Y_{mR}^{n_{mR}}$$

$$= \frac{n_{m1}(n_m-1)!}{n_{m1}(n_{m1}-1)!\,n_{m2}!\cdots n_{mR}!} Y_{m1} Y_{m1}^{n_{m1}-1} Y_{m2}^{n_{m2}} \cdots Y_{mR}^{n_{mR}}$$

$$+ \frac{n_{m2}(n_m-1)!}{n_{m1}!\,n_{m2}(n_{m2}-1)!\cdots n_{mR}!} Y_{m1}^{n_{m1}} Y_{m2} Y_{m2}^{n_{m2}-1} \cdots Y_{mR}^{n_{mR}}$$

$$\vdots$$

$$+ \frac{n_{mR}(n_m-1)!}{n_{m1}!\,n_{m2}!\cdots n_{mR}(n_{mR}-1)!} Y_{m1}^{n_{m1}} Y_{m2}^{n_{m2}} \cdots Y_{mR} Y_{mR}^{n_{mR}-1}$$

$$= \sum_{r=1}^{R} Y_{mr}\, f_m(\underline{n}_R - \underline{1}_r)$$

If any argument of f_m is negative, then the value of the function is defined to be zero.

Lemma: If station m is load dependent and has a FCFS, PS, or

LCFS service discipline, then

$$f_m(\underline{n}_R) = \frac{1}{\mu_m(n_m)} \sum_{r=1}^{R} e_{mr}\, f_m(\underline{n}_R - \underline{1}_r) .$$

Proof:

$$f_m(\underline{n}_R) = \frac{1}{\prod_{j=1}^{n_m} \mu_m(j)} \frac{(n_{m1}+n_{m2}+\cdots+n_{mR})!}{n_{m1}!\,n_{m2}!\cdots n_{mR}!} e_{m1}^{n_{m1}} e_{m2}^{n_{m2}} \cdots e_{mR}^{n_{mR}} \times$$

$$\frac{1}{\mu_m(n_m)} \left[\frac{1}{\prod_{j=1}^{n_m-1} \mu_m(j)} \times \left\{ \frac{n_{m1}(n_m-1)!}{n_{m1}(n_{m1}-1)!\,n_{m2}!\cdots n_{mR}!} e_{m1} e_{m1}^{n_{m1}-1} e_{m2}^{n_{m2}} \cdots e_{mR}^{n_{mR}} \right.\right.$$

$$+ \frac{n_{m2}(n_m-1)!}{n_{m1}!\,n_{m2}(n_{m2}-1)!\cdots n_{mR}!} e_{m1}^{n_{m1}} e_{m2} e_{m2}^{n_{m2}-1} \cdots e_{mR}^{n_{mR}}$$

$$\vdots$$

$$\left.\left. + \frac{n_{mR}(n_m-1)!}{n_{m1}!\,n_{m2}!\cdots n_{mR}(n_{mR}-1)!} e_{m1}^{n_{m1}} e_{m2}^{n_{m2}} \cdots e_{mR} e_{mR}^{n_{mR}-1} \right\} \right]$$

$$= \frac{1}{\mu_m(n_m)} \sum_{r=1}^{R} e_{mr}\, f_m(\underline{n}_R - \underline{1}_r)$$

If any argument of f_m is negative, then the value of the

function is defined to be zero.

Lemma: If station m contains a_m identical servers and has a FCFS service discipline, then

$$f_m(\underline{n}_R) = d_m(\underline{n}_R) \sum_{r=1}^{R} Y_{mr} f'_m(\underline{n}_R-\underline{1}_r)$$

where

$$d_m(\underline{n}_R) = \begin{cases} \dfrac{1}{n_m!} & \text{if } 0 \leqq n_m \leqq a_m \text{ and } n_m = n_{m1}+n_{m2}+\cdots+n_{mR} \\[2ex] \dfrac{1}{a_m! a_m^{n_m-a_m}} & \text{if } n_m \geqq a_m \end{cases}$$

and

$$f'_m(\underline{n}_R) = n_m! \prod_{r=1}^{R} \frac{1}{n_{mr}!} Y_{mr}^{n_{mr}} .$$

where $k_m=k_1+k_2+\cdots+k_R$.

Proof:

$$f_m(\underline{n}_R) = \frac{n_m!}{\prod_{j=1}^{n_m} \mu_m(j)} \prod_{r=1}^{R} \frac{1}{n_{mr}!} e_{mr}^{n_{mr}}$$

$$= \frac{n_m! d_m(\underline{n}_R)}{\mu_{m1}^{n_m}} \prod_{r=1}^{R} \frac{1}{n_{mr}!} e_{mr}^{n_{mr}}$$

$$= d_m(\underline{n}_R)\ n_m!\ \prod_{r=1}^{R} \frac{1}{n_{mr}!} (e_{mr}s_{mr})^{n_{mr}}$$

$$= d_m(\underline{n}_R)\ f'_m(\underline{n}_R)$$

$$= d_m(\underline{n}_R)\ \sum_{r=1}^{R} Y_{mr}\ f'_m(\underline{n}_R-\underline{1}_r) \quad \text{(by Lemma 1)}$$

Eq. 2.9: If station m contains a_m identical servers and has a FCFS queueing discipline, then

$$g_m(\underline{n}_R) = g_{m-1}(\underline{n}_R) + \frac{1}{a_m} \sum_{r=1}^{R} Y_{mr} \left[g_m(\underline{n}_R-\underline{1}_r) + \sum_{\substack{\underline{k}_R=\underline{0}_R \\ k_m \leq a_m-2}}^{\underline{n}_R-\underline{1}_r} \left(\frac{a_m-k_m-1}{k_m+1} \right) f_m(\underline{k}_R) g_{m-1}(\underline{n}_R-\underline{1}_r-\underline{k}_R) \right]$$

Proof:

$$g_m(\underline{n}_R) = \sum_{\underline{k}_R=\underline{0}_R}^{\underline{n}_R} f_m(\underline{k}_R)\ g_{m-1}(\underline{n}_R-\underline{k}_R)$$

$$= g_{m-1}(\underline{n}_R) + \sum_{\underline{k}_R=\underline{0}_R}^{\underline{n}_R} \delta(\underline{k}_R) f_m(\underline{k}_R) g_{m-1}(\underline{n}_R-\underline{k}_R)$$

$$= g_{m-1}(\underline{n}_R) + \sum_{r=1}^{R} Y_{mr}\left[\sum_{\underline{k}_R=\underline{1}_r}^{\underline{n}_R} d_m(\underline{k}_R) f'_m(\underline{k}_R-\underline{1}_r) g_{m-1}(\underline{n}_R-\underline{k}_R)\right]$$

$$= g_{m-1}(\underline{n}_R) + \sum_{r=1}^{R} Y_{mr}\left[\sum_{\underline{k}_R=\underline{0}_R}^{\underline{n}_R-\underline{1}_r} d_m(\underline{k}_R+\underline{1}_r) f'_m(\underline{k}_R) g_{m-1}(\underline{n}_R-\underline{1}_r-\underline{k}_R)\right]$$

$$= g_{m-1}(\underline{n}_R) + \sum_{r=1}^{R} Y_{mr}\Big[\sum_{\substack{\underline{k}_R=\underline{0}_R \\ k_m \leqq a_m-1}}^{\underline{n}_R-\underline{1}_r} \frac{1}{k_m+1} f_m(\underline{k}_R) g_{m-1}(\underline{n}_R-\underline{1}_r-\underline{k}_R)$$

$$+ \sum_{\substack{\underline{k}_R=\underline{0}_R \\ k_m \geqq a_m}}^{\underline{n}_R-\underline{1}_r} \frac{1}{a_m} f_m(\underline{k}_R) g_{m-1}(\underline{n}_R-\underline{1}_r-\underline{k}_R)\Big]$$

But,

$$\frac{g_m(\underline{n}_R-\underline{1}_r)}{a_m} = \sum_{\substack{\underline{k}_R=\underline{0}_R \\ k_m \leqq a_m-1}}^{\underline{n}_R-\underline{1}_r} \frac{1}{a_m} f_m(\underline{k}_R) g_{m-1}(\underline{n}_R-\underline{1}_r-\underline{k}_R)$$

$$\sum_{\substack{\underline{k}_R=\underline{0}_R \\ k_m \geqq a_m}}^{\underline{n}_R-\underline{1}_r} \frac{1}{a_m} f_m(\underline{k}_R) g_{m-1}(\underline{n}_R-\underline{1}_r-\underline{k}_R)$$

Thus,

$$g_m(\underline{n}_R) = g_{m-1}(\underline{n}_R) + \frac{1}{a_m} \sum_{r=1}^{R} Y_{mr}\ \Big[g_m(\underline{n}_R-\underline{1}_r)$$

$$+ \sum_{\substack{\underline{k}_R=\underline{0}_R \\ k_m \leqq a_m-2}}^{\underline{n}_R-\underline{1}_r} \left(\frac{a_m-k_m-1}{k_m+1}\right) f_m(\underline{k}_R) g_{m-1}(\underline{n}_R-\underline{1}_r-\underline{k}_R)\Big]$$

The last summation denotes R nested sums. The rth sum need only range to $\min(n_r, a_m-2)$.

<u>Eq. 2.11</u>: If station i is load independent, then

$$g_m^{i}(\underline{n}_R) = G(\underline{n}_R) - \sum_{r=1}^{R} Y_{ir}\ G(\underline{n}_R-\underline{1}_r)$$

Proof:

$$g^{i}_{M}(\underline{n}_R) = G(n_R) - \sum_{\underline{k}_R=\underline{0}_R}^{\underline{n}_R} \delta(\underline{k}_R)\, f_i(\underline{k}_R)\, g^{i}_{M}(\underline{n}_R-\underline{k}_R)$$

$$= G(\underline{n}_R) - \sum_{r=1}^{R} Y_{ir} \left[\sum_{\underline{k}_R=\underline{0}_R}^{\underline{n}_R} \delta(\underline{k}_R) f_i(\underline{k}_R-\underline{1}_r) g^{i}_{M}(\underline{n}_R-\underline{k}_R) \right]$$

$$= G(\underline{n}_R) - \sum_{r=1}^{R} Y_{ir} \left[\sum_{\underline{k}_R=\underline{1}_r}^{\underline{n}_R} f_i(\underline{k}_R-\underline{1}_r)\, g^{i}_{M}(\underline{n}_R-\underline{k}_R) \right]$$

$$= G(\underline{n}_R) - \sum_{r=1}^{R} Y_{ir} \left[\sum_{\underline{k}_R=\underline{0}_R}^{\underline{n}_R-\underline{1}_r} f_i(\underline{k}_R)\, g^{i}_{M}(\underline{n}_R-\underline{1}_r-\underline{k}_r) \right]$$

$$= G(\underline{n}_R) - \sum_{r=1}^{R} Y_{ir}\, G(\underline{n}_R-\underline{1}_r)$$

Eq. 2.12: If station i contains a_i identical servers, then

$$g^{i}_{M}(\underline{n}_R) = G(\underline{n}_R) - \frac{1}{a_i} \sum_{r=1}^{R} Y_{ir} \Big[G(\underline{n}_R-\underline{1}_r) + \sum_{\substack{\underline{k}_R=\underline{0}_R \\ k_i \leqq a_i-2}}^{\underline{n}_R-\underline{1}_r} \left(\frac{a_i-k_i-1}{k_i+1} \right) f_i(\underline{k}_R)\, g^{i}_{M}(\underline{n}_R-\underline{1}_r-\underline{k}_R) \Big]$$

Proof: Similar to proof of equation 2.9.

<u>Lemma 2</u>: For any FCFS, PS, LCFS, or IS station i ($1 \leqq i \leqq M$) and for any job class r ($1 \leqq r \leqq R$), if $n_{ir} \neq 0$, then

$$f_i(\underline{n}_i) = \begin{cases} e_{ir} \dfrac{n_i \; f_i(\underline{n}_i - \underline{1}_r)}{n_{ir}\mu_i(n_i)} & \text{if station i has a FCFS, PS, or LCFS service discipline} \\ e_{ir} \dfrac{s_{ir}}{n_i} f_i(\underline{n}_i - \underline{1}_r) & \text{if station i has an IS service discipline} \end{cases}$$

<u>Proof</u>:

For a FCFS, PS, or LCFS station:

$$f_i(\underline{n}_i) = \frac{n_i!}{\prod_{j=1}^{n_i} \mu_i(j)} \prod_{s=1}^{R} \frac{1}{n_{is}!} e_{is}^{n_{is}}$$

$$= \frac{n_i(n_i-1)!}{\mu_i(n_i) \prod_{j=1}^{n_i-1} \mu_i(j)} \frac{e_{ir} e_{ir}^{n_{ir}-1}}{n_{ir}(n_{ir}-1)!} \prod_{\substack{s=1 \\ s \neq r}}^{R} \frac{1}{n_{is}!} e_{is}^{n_{is}}$$

$$= \frac{n_i}{n_{ir}} \frac{e_{ir}}{\mu_i(n_i)} f_i(\underline{n}_i - \underline{1}_r)$$

For an IS station:

$$f_i(\underline{n}_i) = \prod_{j=1}^{R} \frac{1}{n_{ij}!} (e_{ij} s_{ij})^{n_{ij}}$$

$$= \frac{1}{n_{ir}(n_{ir}-1)!} (e_{ir}s_{ir})(e_{ir}s_{ir})^{n_{ir}-1} \prod_{\substack{j=1 \\ j\neq r}}^{R} (e_{ij}s_{ij})^{n_{ij}}$$

$$= \frac{e_{ir}s_{ir}}{n_{ir}} f_i(\underline{n}_i-\underline{1}_r)$$

<u>Proof of multiple class throughput formula</u>:

$$X_{ir}(\underline{N}_R) = e_{ir} \frac{G(\underline{N}_R-\underline{1}_r)}{G(\underline{N}_R)}$$

For a FCFS, PS, or LCFS station:

$$X_{ir}(\underline{N}_R) = \sum_{\underline{n}_R=\underline{1}_r}^{\underline{N}_R} p_i(\underline{n}_R,\underline{N}_R) \frac{n_{ir}}{n_i} \mu_i(n_i)$$

$$= \sum_{\underline{n}_R=\underline{1}_r}^{\underline{N}_R} \frac{f_i(\underline{n}_R)}{G(\underline{N}_R)} g_M^i(\underline{N}_R-\underline{n}_R) \frac{n_{ir}}{n_i} \mu_i(n_i)$$

$$= e_{ir} \sum_{\underline{n}_R=\underline{1}_r}^{\underline{N}_R} \frac{f_i(\underline{n}_R-\underline{1}_r)}{G(\underline{N}_R)} g_M^i(\underline{N}_R-\underline{n}_R)$$

$$= e_{ir} \sum_{\underline{n}_R=\underline{0}_R}^{\underline{N}_R-\underline{1}_r} \frac{f_i(\underline{n}_R)}{G(\underline{N}_R)} g_M^i(\underline{N}_R-\underline{1}_r-\underline{n}_R)$$

$$= e_{ir} \frac{G(\underline{N}_R-\underline{1}_r)}{G(\underline{N}_R)}$$

For an IS station:

$$X_{ir}(\underline{N}_R) = \sum_{\underline{n}_R=\underline{1}_r}^{\underline{N}_R} p_i(\underline{n}_R,\underline{N}_R)\,\frac{n_{ir}}{s_{ir}}$$

$$= \sum_{\underline{n}_R=\underline{1}_r}^{\underline{N}_R} \frac{f_i(\underline{n}_R)}{G(\underline{N}_R)}\, g_M^{i}(\underline{N}_R-\underline{n}_R)\,\frac{n_{ir}}{s_{ir}}$$

$$= e_{ir} \sum_{\underline{n}_R=\underline{1}_r}^{\underline{N}_R} \frac{f_i(\underline{n}_R-\underline{1}_r)}{G(\underline{N}_R)}\, g_M^{i}(\underline{N}_R-\underline{n}_R)$$

$$= e_{ir} \sum_{\underline{n}_R=\underline{0}_R}^{\underline{N}_R-\underline{1}_r} \frac{f_i(\underline{n}_R)}{G(\underline{N}_R)}\, g_M^{i}(\underline{N}_R-\underline{1}_r-\underline{n}_R)$$

$$= e_{ir}\,\frac{G(\underline{N}_R-\underline{1}_r)}{G(\underline{N}_R)}$$

<u>Lemma</u>: If station i is load independent and has a FCFS, PS, or LCFS queueing discipline, then the marginal queue length distribution can be written as

$$p_i(\underline{n}_R,\underline{N}_R) = \sum_{r=1}^{R} s_{ir}\, X_{ir}(\underline{N}_R)\, p_i(\underline{n}_R-\underline{1}_r,\underline{N}_R-\underline{1}_r)$$

<u>Proof</u>:

$$p_i(\underline{n}_R,\underline{N}_R) = \frac{f_i(\underline{n}_R)}{G(\underline{N}_R)}\, g_M^{i}(\underline{N}_R-\underline{n}_R)$$

$$= \sum_{r=1}^{R} Y_{ir}\ f_i(\underline{n}_R-\underline{1}_r)\ \frac{g_M^{i}(\underline{N}_R-\underline{n}_R)}{G(\underline{N}_R)} \qquad \text{by Lemma 1}$$

$$= \sum_{r=1}^{R} s_{ir}\ e_{ir}\ \frac{G(\underline{N}_R-\underline{1}_r)}{G(\underline{N}_R)}\ f_i(\underline{n}_R-\underline{1}_r)\ \frac{g_M^{i}(\underline{N}_R-\underline{n}_r)}{G(\underline{N}_R-\underline{1}_r)}$$

$$= \sum_{r=1}^{R} s_{ir}\ X_{ir}(\underline{N}_R)\ p_i(\underline{n}_R-\underline{1}_r,\underline{N}_R-\underline{1}_r)$$

Lemma: If station i is load independent and has a FCFS, PS, or LCFS queueing discipline, then the marginal queue size distribution can be written

$$P_i(n,\underline{N}_R) = \sum_{r=1}^{R} s_{ir}\ X_{ir}(\underline{N}_R)\ P_i(n-1,\underline{N}_R-\underline{1}_r) \qquad \text{for } n>0.$$

Proof:

$$P_i(n,\underline{N}_R) = \sum_{\substack{\underline{n}_R=\underline{0}_R \\ n_i=n}}^{\underline{N}_R} p_i(\underline{n}_R,\underline{N}_R) \qquad \text{where } n_i=n_1+n_2+\cdots+n_R$$

$$= \sum_{r=1}^{R} s_{ir}\ X_{ir}(\underline{N}_R) \sum_{\substack{\underline{n}_R=\underline{0}_R \\ n_i=n}}^{\underline{N}_R} p_i(\underline{n}_R-\underline{1}_r,\underline{N}_R-\underline{1}_r)$$

$$= \sum_{r=1}^{R} s_{ir}\ X_{ir}(\underline{N}_R)\ P_i(n-1,\underline{N}_R-\underline{1}_r)$$

Multiple class load independent global mean queue length

$$\bar{n}_i(\underline{N}_R) = \sum_{r=1}^{R} U_{ir}(\underline{N}_R)\ (1+\bar{n}_i(\underline{N}_R-\underline{1}_r))$$

Proof:

$$\bar{n}_i(\underline{N}_R) = \sum_{n=1}^{N} n\ P_i(n,\underline{N}_R)$$

$$= \sum_{n=1}^{N} n \sum_{r=1}^{R} s_{ir}\ X_{ir}(\underline{N}_R)\ P_i(n-1,\underline{N}_R-\underline{1}_r)$$

$$= \sum_{r=1}^{R} s_{ir}\ X_{ir}(\underline{N}_R) \sum_{n=1}^{N} n\ P_i(n-1,\underline{N}_R-\underline{1}_r)$$

$$= \sum_{r=1}^{R} U_{ir}(\underline{N}_R) \sum_{n=0}^{N-1} (n+1)\ P_i(n,\underline{N}_R-\underline{1}_r)$$

$$= \sum_{r=1}^{R} U_{ir}(\underline{N}_R)\ \left(\sum_{n=0}^{N-1} nP_i(n,\underline{N}_R-\underline{1}_r) + \sum_{n=0}^{N-1} P_i(n,\underline{N}_R-\underline{1}_r) \right)$$

$$= \sum_{r=1}^{R} U_{ir}(\underline{N}_R)\ (1 + \bar{n}_i(\underline{N}_R-\underline{1}_r))$$

Multiple class load independent mean queue length:

$$\bar{n}_{ir}(\underline{N}_R) = U_{ir}(\underline{N}_R)\ (1 + \bar{n}_i(\underline{N}_R-\underline{1}_r))$$

Proof:

$$\bar{n}_{ir}(\underline{N}_R) = \sum_{\underline{n}_R=\underline{1}_r}^{\underline{N}_R} p_i(\underline{n}_R,\underline{N}_R)\ n_{ir}$$

$$= \sum_{\underline{n}_R=\underline{1}_r}^{\underline{N}_R} n_{ir} \frac{f_i(\underline{n}_R)}{G(\underline{N}_R)} g^i_M(\underline{N}_R-\underline{n}_R)$$

$$= \sum_{\underline{n}_R=\underline{1}_r}^{\underline{N}_R} n_{ir} \frac{n_i}{n_{ir}} s_{ir} f_i(\underline{n}_R-\underline{1}_r) e_{ir} \frac{G(\underline{N}_R-\underline{1}_r)}{G(\underline{N}_R)} \frac{g^i_M(\underline{N}_R-\underline{n}_R)}{G(\underline{N}_R-\underline{1}_r)}$$

$$= s_{ir} X_{ir}(\underline{N}_R) \sum_{\underline{n}_R=\underline{1}_r}^{\underline{N}_R} n_i \frac{f_i(\underline{n}_R-\underline{1}_r)}{G(\underline{N}_R-\underline{1}_r)} g^i_M(\underline{N}_R-\underline{1}_r-(\underline{n}_R-\underline{1}_r))$$

$$= U_{ir}(\underline{N}_R) \sum_{\underline{n}_R=\underline{1}_r}^{\underline{N}_R} n_i p_i(\underline{n}_R-\underline{1}_r,\underline{N}_R-\underline{1}_r)$$

$$= U_{ir}(\underline{N}_R) \sum_{\underline{n}_R=\underline{0}_r}^{\underline{N}_R-\underline{1}_r} (n_i+1) p_i(\underline{n}_R,\underline{N}_R-\underline{1}_r)$$

$$= U_{ir}(\underline{N}_R) \left(\sum_{\underline{n}_R=\underline{0}_R}^{\underline{N}_R-\underline{1}_r} p_i(\underline{n}_R,\underline{N}_R-\underline{1}_r) + \sum_{\underline{n}_R=\underline{0}_R}^{\underline{N}_R-\underline{1}_r} n_i p_i(\underline{n}_R,\underline{N}_R-\underline{1}_r) \right)$$

$$= U_{ir}(\underline{N}_R) (1 + \bar{n}_i(\underline{N}_R-\underline{1}_r))$$

<u>Multinomial sum</u>:

$$\left(\sum_{r\in S} x_r \right)^k = \sum_{\sum_{r\in S} n_r=k} k! \prod_{r\in S} \frac{x_r^{n_r}}{n_r!}$$

for some index set S.

The multinomial sum (or theorem) includes the binomial theorem as a special case. For our purposes the index set S is usually a subset of the set of all job classes {1,2,...,R}. To aid in understanding the above expression, consider a simple example with the index set S being {1,2,3}. Then from the multinomial sum

$$(x_1+x_2+x_3)^k = \sum_{n_1+n_2+n_3=k} k! \frac{x_1^{n_1} \; x_2^{n_2} \; x_3^{n_3}}{n_1! \; n_2! \; n_3!}$$

where the sum is taken over all nonnegative integers n_1, n_2, n_3 for which $n_1+n_2+n_3=k$.

Lemma 3: The expression

$$\sum_{\underline{k}_U=\underline{0}_U}^{\underline{n}_U} [\sum_{\underline{n}_i=\underline{k}_U} f_i(\underline{n}_i) \; \varphi(\underline{k}_U)]$$

where φ is a function of the index variables $k_1,k_2,\ldots,k_U$ reduces to

$$\sum_{\underline{k}_U=\underline{0}_U}^{\underline{n}_U} fp_i(\underline{k}_U) \; \varphi(\underline{k}_U)$$

where

$$fp_i(\underline{k}_U) = \begin{cases} \dfrac{k_i!}{\prod_{k=1}^{k_i} \mu_i(k)} \prod_{s=1}^{U} \dfrac{1}{k_s!} \left[\sum_{r \in EC_s} (e_{ir} s_{ir}) \right]^{k_s} \\[2ex] k_i! \prod_{s=1}^{U} \dfrac{1}{k_s!} \left[\sum_{r \in EC_s} (e_{ir} s_{ir}) \right]^{k_s} \\[2ex] \prod_{s=1}^{U} \dfrac{1}{k_s!} \left[\sum_{r \in EC_s} (e_{ir} s_{ir}) \right]^{k_s} \end{cases}$$

and $k_i = \sum_{s=1}^{U} k_s$.

<u>Proof</u>:

$$\sum_{\underline{k}_U=\underline{0}_U}^{\underline{n}_U} \left[\sum_{\underline{n}_i=\underline{k}_U} f_i(\underline{n}_i)\, \varphi(\underline{k}_U) \right]$$

$$= \sum_{\underline{k}_U=\underline{0}_U}^{\underline{n}_U} \varphi(\underline{k}_U) \sum_{\underline{n}_i=\underline{k}_U} f_i(\underline{n}_i)$$

The innermost summation can be simplified by working it into the form of a multinomial sum:

$$\sum_{\underline{n}_i=\underline{k}_U} f_i(\underline{n}_i)$$

$$= \sum_{\substack{\underline{n}_i \in \{(n_{i1}, \ldots, n_{ir}) \mid \\ \sum_{r \in EC_l} n_{ir} = k_l,\ l=1,2,\ldots,U\}}} \prod_{r=1}^{R} \frac{1}{n_{iR}!} (e_{ir} s_{ir})^{n_{ir}}$$

$$= \sum_{\underline{n}_i \in \{(n_{i1},\ldots,n_{iR}) \mid \sum_{r\in EC_l} n_{ir}=k_l,\, l=1,2,\ldots,U\}} \prod_{s=1}^{U} \prod_{r\in EC_s} \frac{1}{n_{ir}!} (e_{ir}s_{ir})^{n_{ir}}$$

$$= \prod_{s=1}^{U} \frac{1}{k_s!} \sum_{\substack{\sum_{r\in EC_l} n_{ir}=k_l,\\ l=1,2,\ldots,U}} k_s! \prod_{r\in EC_s} \frac{1}{n_{ir}!} (e_{ir}s_{ir})^{n_{ir}}$$

$$= \prod_{s=1}^{U} \frac{1}{k_s!} \Big[\sum_{r\in EC_s} (e_{ir}s_{ir})\Big]^{k_s}$$

<u>Eq. 2.18</u>:

$$X_{ir}(\underline{N}_U) = e_{ir} \frac{G(\underline{N}_U-\underline{1}_q)}{G(\underline{N}_U)}$$

where q is the equivalence class of job class r.

<u>Proof</u>:

For a FCFS, PS, or LCFS station:

$$X_{ir}(\underline{N}_U) = \sum_{\underline{n}_U=\underline{1}_q}^{\underline{N}_U} p_i(\underline{n}_R,\underline{N}_U) \frac{n_{ir}}{n_i} \mu_i(n_i)$$

$$= \sum_{\underline{n}_U=\underline{1}_q}^{\underline{N}_U} \frac{f_i(\underline{n}_R)}{G(\underline{N}_U)} g_M^i(\underline{N}_U-\underline{k}_U) \frac{n_{ir}}{n_i} \mu_i(n_i)$$

$$= e_{ir} \sum_{\underline{n}_U=\underline{1}_q}^{\underline{N}_U} \frac{f_i(\underline{n}_R-\underline{1}_q)}{G(\underline{N}_U)} g_M^i(\underline{N}_U-\underline{k}_U)$$

$$= e_{ir} \sum_{\underline{n}_U=\underline{0}_U}^{\underline{N}_U-\underline{1}_q} \frac{f_i(\underline{n}_R)}{G(\underline{N}_U)} g_M^i(\underline{N}_U-\underline{1}_q-\underline{k}_U)$$

$$= e_{ir} \frac{G(\underline{N}_U-\underline{1}_q)}{G(\underline{N}_U)}$$

For an IS station:

$$X_{ir}(\underline{N}_U) = \sum_{\underline{n}_U=\underline{1}_q}^{\underline{N}_U} p_i(\underline{n}_R,\underline{N}_U) \frac{n_{ir}}{s_{ir}}$$

$$= \sum_{\underline{n}_U=\underline{1}_q}^{\underline{N}_U} \frac{f_i(\underline{n}_R)}{G(\underline{N}_U)} g_M^i(\underline{N}_U-\underline{k}_U) \frac{n_{ir}}{s_{ir}}$$

$$= e_{ir} \sum_{\underline{n}_U=\underline{1}_q}^{\underline{N}_U} \frac{f_i(\underline{n}_R-\underline{1}_q)}{G(\underline{N}_U)} g_M^i(\underline{N}_U-\underline{k}_U)$$

$$= e_{ir} \sum_{\underline{n}_U=\underline{0}_U}^{\underline{N}_U-\underline{1}_q} \frac{f_i(\underline{n}_R)}{G(\underline{N}_U)} g_M^i(\underline{N}_U-\underline{1}_q-\underline{k}_U)$$

$$= e_{ir} \frac{G(\underline{N}_U-\underline{1}_q)}{G(\underline{N}_U)}$$

Lemma: If station i is load independent and has a FCFS, PS, or LCFS queueing discipline, then the marginal queue length distribution can be written as

$$p_i(\underline{n}_R,\underline{N}_U) = \sum_{r=1}^{R} s_{ir} X_{ir}(\underline{N}_U)\, p_i(\underline{n}_R-\underline{1}_r,\underline{N}_U-\underline{1}_q)$$

where q is the equivalence class of job class r.

Proof:

$$p_i(\underline{n}_R,\underline{N}_U) = \frac{f_i(\underline{n}_R)}{G(\underline{N}_U)}\, g_M^i(\underline{N}_U-\underline{k}_U)$$

(where $k_U=(k_1,k_2,\ldots,k_U)$ and $k_l=\sum_{r\in EC_l} n_{ir}$ for $l=1,2,\ldots,U$)

$$= \sum_{r=1}^{R} Y_{ir}\, f_i(\underline{n}_R-\underline{1}_r)\, \frac{g_M^i(\underline{N}_U-\underline{k}_U)}{G(\underline{N}_U)} \quad \text{by Lemma 1}$$

$$= \sum_{r=1}^{R} s_{ir}\, e_{ir}\, \frac{G(\underline{N}_U-\underline{1}_q)}{G(\underline{N}_U)}\, f_i(\underline{n}_R-\underline{1}_r)\, \frac{g_M^i(\underline{N}_U-\underline{k}_U)}{G(\underline{N}_U-\underline{1}_q)}$$

$$= \sum_{r=1}^{R} s_{ir} X_{ir}(\underline{N}_U)\, p_i(\underline{n}_R-\underline{1}_r,\underline{N}_U-\underline{1}_q)$$

Lemma: If station i is load independent and has a FCFS, PS, or LCFS queueing discipline, then the marginal queue size distribution can be written

$$P_i(n,\underline{N}_U) = \sum_{r=1}^{R} s_{ir}\ X_{ir}(\underline{N}_U)\ P_i(n-1,\underline{N}_U-\underline{1}q) \quad \text{for } n>0.$$

where q is the equivalence class of job class r.

Proof:

$$P_i(n,\underline{N}_U) = \sum_{\substack{\underline{n}_U=\underline{0}_U \\ n_i=n}}^{\underline{N}_U} p_i(\underline{n}_R,\underline{N}_U) \qquad \text{where } n_i=n_1+n_2+\cdots+n_R$$

$$= \sum_{r=1}^{R} s_{ir}\ X_{ir}(\underline{N}_U) \sum_{\substack{\underline{n}_R=\underline{0}_U \\ n_i=n}}^{\underline{N}_U} p_i(\underline{n}_R-\underline{1}_r,\underline{N}_U-\underline{1}_q)$$

$$= \sum_{r=1}^{R} s_{ir}\ X_{ir}(\underline{N}_U)\ P_i(n-1,\underline{N}_U-\underline{1}_q)$$

Multiple class load independent global mean queue length

$$\bar{n}_i(\underline{N}_U) = \sum_{r=1}^{R} U_{ir}(\underline{N}_U)\ (1+\bar{n}_i(\underline{N}_U-\underline{1}_q))$$

where q is equivalence class of job class r.

Proof:

$$\bar{n}_i(\underline{N}_U) = \sum_{n=1}^{N} n\ P_i(n,\underline{N}_U)$$

$$= \sum_{n=1}^{N} n \sum_{r=1}^{R} s_{ir} X_{ir}(\underline{N}_U) P_i(n-1, \underline{N}_U - \underline{1}_q)$$

$$= \sum_{r=1}^{R} s_{ir} X_{ir}(\underline{N}_U) \sum_{n=1}^{N} n P_i(n-1, \underline{N}_U - \underline{1}_q)$$

$$= \sum_{r=1}^{R} U_{ir}(\underline{N}_U) \sum_{n=0}^{N-1} (n+1) P_i(n, \underline{N}_U - \underline{1}_q)$$

$$= \sum_{r=1}^{R} U_{ir}(\underline{N}_U) \left(\sum_{n=0}^{N-1} n P_i(n, \underline{N}_U - \underline{1}_q) + \sum_{n=0}^{N-1} P_i(n, \underline{N}_U - \underline{1}_q) \right)$$

$$= \sum_{r=1}^{R} U_{ir}(\underline{N}_U) (1 + \bar{n}_i(\underline{N}_U - \underline{1}_q))$$